Collins

Student Support Materials for Edexcel A Level Maths Core 2

Authors: John Berry and Sue Langham

William Collins' dream of knowledge for all began with the publication of his first book in 1819. A self-educated mill worker, he not only enriched millions of lives, but also founded a flourishing publishing house. Today, staying true to this spirit, Collins books are packed with inspiration, innovation and practical expertise. They place you at the centre of a world of possibility and give you exactly what you need to explore it.

Collins. Freedom to teach.

Published by Collins
An imprint of HarperCollins *Publishers*
77–85 Fulham Palace Road
Hammersmith
London
W6 8JB

10 9 8 7 6 5 4 3 2 1

ISBN-13: 978-0-00-747602-2

British Library Cataloguing in Publication Data. A Catalogue record for this publication is available from the British Library.

Commissioned by Lindsey Charles and Emma Braithwaite
Project managed by Lindsey Charles
Edited and proofread by Susan Gardner
Reviewed by Stewart Townend
Design and typesetting by Jouve India Private Limited
Illustrations by Ann Paganuzzi
Index compiled by Michael Forder
Cover design by Angela English
Production by Simon Moore

Printed and bound in Spain by Graficas Estella

This material has been endorsed by Edexcel and offers high quality support for the delivery of Edexcel qualifications.

Edexcel endorsement does not mean that this material is essential to achieve any Edexcel qualification, nor does it mean that this is the only suitable material available to support any Edexcel qualification. No endorsed material will be used verbatim in setting any Edexcel examination and any resource lists produced by Edexcel shall include this and other appropriate texts. While this material has been through an Edexcel quality assurance process, all responsibility for the content remains with the publisher.

Copies of official specifications for all Edexcel qualifications may be found on the Edexcel website - www.edexcel.com

Browse the complete Collins catalogue at: www.collinseducation.com

Acknowledgements
The publishers wish to thank the following for permission to reproduce photographs. Every effort has been made to trace copyright holders and to obtain their permission for the use of copyright material. The publishers will gladly receive any information enabling them to rectify any error or omission at the first opportunity.

Cover image: Architectural abstract - fragment of modern glass roof © travelif | istockphoto.com

Welcome to Collins Student Support Materials for Edexcel A level Mathematics. This page introduces you to the key features of the book which will help you to succeed in your examinations and to enjoy your maths course.

The chapters are organised by the main sections within the specification for easy reference. Each one gives a succinct explanation of the key ideas you need to know.

Examples and answers

After ideas have been explained the worked examples in the green boxes demonstrate how to use them to solve mathematical problems.

Method notes

These appear alongside some of the examples to give more detailed help and advice about working out the answers.

Essential ideas

These are other ideas which you will find useful or need to recall from previous study.

Exam tips

These tell you what you will be expected to do, or not to do, in the examination.

Stop and think

The stop and think sections present problems and questions to help you reflect on what you have just been reading. They are not straightforward practice questions - you have to think carefully to answer them!

Practice examination section

At the end of the book you will find a section of practice examination questions which help you prepare for the ones in the examination itself. Answers with full workings out are provided to all the examination questions so that you can see exactly where you are getting things wrong or right!

Notation and formulae

The notation and formulae used in this examination module are listed at the end of the book just before the index for easy reference. The formulae list shows both what you need to know and what you will be given in the exam.

Contents

In Core 1 you were introduced to polynomials and how to add, subtract and multiply them. The **degree** of a polynomial is the index of the highest power of x in the expression with a non-zero coefficient:

$9x^7 - 2x^5 + 12x^4 + x^2 - 17x + 3$ is a polynomial of degree 7

$2x^3 + 5x^2 - 7x - 15$ is a polynomial of degree 3

$x^4 + 11x^3 - 2x^2 - 5x + 3$ is a polynomial of degree 4

$3x^5 - 5x^4 + 12x^3 + x^2 - 7x + 13$ is a polynomial of degree 5

Essential notes

Some polynomials are given special names:

'a cubic' is of degree 3

'a quartic' is of degree 4

'a quintic' is of degree 5

Linear, quadratic and cubic functions are examples of **polynomial functions** in which each term is a multiple of a power of x.

In this chapter we explore the division of polynomials and methods for finding factors of polynomials.

Algebraic division of polynomials

Example

Simplify these fractions.

a) $\dfrac{2x^3 + 6x^2 - 14x}{2x}$

b) $\dfrac{x^2 - 8x + 15}{(x - 3)}$

c) $\dfrac{x^2 + x - 6}{(x^2 - 4)}$

Answer

a) $\dfrac{2x^3 + 6x^2 - 14x}{2x} = \dfrac{2x^3}{2x} + \dfrac{6x^2}{2x} - \dfrac{14x}{2x} = x^2 + 3x - 7$

b) $\dfrac{x^2 - 8x + 15}{(x - 3)} = \dfrac{(x - 3)(x - 5)}{(x - 3)} = x - 5$

c) $\dfrac{x^2 + x - 6}{(x^2 - 4)} = \dfrac{(x - 2)(x + 3)}{(x - 2)(x + 2)} = \dfrac{x + 3}{x + 2}$

Method notes

a) Divide each term in the numerator by $2x$.

b) First factorise the numerator then cancel out the common factor $(x - 3)$ in the numerator and the denominator.

c) Factorise the numerator and the denominator then cancel out the common factor $(x - 2)$ in the numerator and the denominator.

In these examples we were able to factorise the numerator and denominator of the algebraic fractions. The following example shows an alternative method of simplifying algebraic fractions where we use 'long division' to divide a polynomial by a **linear** function.

Essential notes

A linear function is of the form $(x + p)$ or $(x - p)$ where p is a positive constant.

Example

Divide $3x^3 + 14x^2 - 3x + 10$ by $(x + 5)$ and hence write $3x^3 + 14x^2 - 3x + 10$ as a product of two factors.

Answer

$$\begin{array}{rl}
 & 3x^2 \\
(x+5) & \overline{)3x^3 + 14x^2 - 3x + 10} \\
 & \underline{3x^3 + 15x^2} \\
 & \qquad -x^2 - 3x + 10
\end{array}$$

$$\begin{array}{rl}
 & 3x^2 - x \\
(x+5) & \overline{)3x^3 + 14x^2 - 3x + 10} \\
 & \underline{3x^3 + 15x^2} \\
 & \qquad -x^2 - 3x + 10 \\
 & \qquad \underline{-x^2 - 5x} \\
 & \qquad\qquad 2x + 10
\end{array}$$

$$\begin{array}{rl}
 & 3x^2 - x + 2 \longleftarrow \text{quotient} \\
(x+5) & \overline{)3x^3 + 14x^2 - 3x + 10} \\
 & \underline{3x^3 + 15x^2} \\
 & \qquad -x^2 - 3x + 10 \\
 & \qquad \underline{-x^2 - 5x} \\
 & \qquad\qquad 2x + 10 \\
 & \qquad\qquad \underline{2x + 10} \\
 & \qquad\qquad\qquad 0 \longleftarrow \text{remainder}
\end{array}$$

Method notes

$(x + 5)$ is the linear divisor.

Step 1: Divide $3x^3$ by x from the linear divisor $= 3x^2$

Write this above the 'bridge'.

Step 2: Multiply $(x + 5)$ by the result from step $1 = 3x^3 + 15x^2$.

Step 3: Subtract the result from step 2 from the polynomial under the 'bridge' $= -x^2 - 3x + 10$

Step 4: Divide $-x^2$ from step 3 by x from the linear divisor $= -x$ and write this above the 'bridge'.

Step 5: Multiply $(x + 5)$ by the result from step $4 = -x^2 - 5x$ and **subtract** this result from the result in step $3 = 2x + 10$

Step 6: Divide $2x$ from step 5 by x from the linear divisor = 2 and write this above the 'bridge'.

Step 7: Multiply $(x + 5)$ by the result from step $6 = 2x + 10$ and **subtract** this result from $2x + 10$

Once the answer from the subtraction is of lower degree than the divisor you stop dividing out. The answer above the bridge, in this example $(3x^2 - x + 2)$, is called the **quotient**. Anything left when you stop the dividing out is called the **remainder**.

In this example the final step gave 0 so there is no remainder and we say that $(x + 5)$ is a **factor** of $3x^3 + 14x^2 - 3x + 10$ therefore we can write $3x^3 + 14x^2 - 3x + 10 = (x + 5)(3x^2 - x + 2)$ which is a product of two factors.

There are many steps involved in algebraic division but they follow a clear pattern which is summarised as follows.

- Divide the first term in the polynomial by the first term in the divisor, multiply the result by the whole linear divisor then subtract this from the original polynomial.
- Divide the first term in the resulting polynomial by the first term in the divisor, multiply the result by the whole linear divisor and then subtract this from the first of the resulting polynomials.
- Continue this process until the polynomial answer from any subtraction is of lower degree than the divisor.

Essential notes

In arithmetic we say that 2 is a **factor** of 16 because we can write $16 = 2 \times 8$ or we can say that if we divide 16 by 2 there is **no remainder**.

Example

Divide $2x^3 - x^2 - 10x + 11$ by $(x - 2)$.

Answer

$$
\begin{array}{rl}
 & 2x^2 + 3x - 4 \quad \leftarrow \boxed{\text{quotient}} \\
(x-2) & \overline{)2x^3 - x^2 - 10x + 11} \\
 & \underline{2x^3 - 4x^2} \\
 & \quad 3x^2 - 10x + 11 \\
 & \quad \underline{3x^2 - 6x} \\
 & \qquad -4x + 11 \\
 & \qquad \underline{-4x + 8} \\
 & \qquad\qquad 3 \quad \leftarrow \boxed{\text{remainder}}
\end{array}
$$

$(2x^3 - x^2 - 10x + 11)$ divided by $(x - 2)$ has quotient $(2x^2 + 3x - 4)$ and remainder 3. In this example $(x - 2)$ is NOT a factor of $(2x^3 - x^2 - 10x + 11)$

Method notes

Follow the method of the previous example.

Divide $2x^3$ by x and write this above the 'bridge'.

Multiply $2x^2$ by $(x - 2)$ and **subtract** this from under the bridge

Divide $3x^2$ by x and write this above the 'bridge'.

Multiply $3x$ by $(x - 2)$ and **subtract** this from under the bridge.

Divide $-4x$ by x to give -4 and write this above the bridge.

Multiply -4 by $(x - 2)$ and **subtract** this from under the bridge.

3 is the remainder.

The quotient is the polynomial above the bridge, $2x^2 + 3x - 4$

The next example follows the same method.

Example

Show that $(x + 1)$ is a factor of $(x^3 - 7x^2 - x + 7)$.

Hence write $(x^3 - 7x^2 - x + 7)$ as a product of three linear factors.

Answer

$$
\begin{array}{rl}
 & x^2 - 8x + 7 \quad \leftarrow \boxed{\text{quotient}} \\
(x+1) & \overline{)x^3 - 7x^2 - x + 7} \\
 & \underline{x^3 + x^2} \\
 & \quad -8x^2 - x + 7 \\
 & \quad \underline{-8x^2 - 8x} \\
 & \qquad 7x + 7 \\
 & \qquad \underline{7x + 7} \\
 & \qquad\quad 0 \quad \leftarrow \boxed{\text{remainder} = 0}
\end{array}
$$

As there is no remainder $(x + 1)$ is a factor of $(x^3 - 7x^2 - x + 7)$. So:

$$(x^3 - 7x^2 - x + 7) = (x + 1)\ (x^2 - 8x + 7).$$

The quadratic polynomial $(x^2 - 8x + 7)$ can then be factorised

$$(x^2 - 8x + 7) = (x - 7)(x - 1)$$

Therefore

$$(x^3 - 7x^2 - x + 7) = (x + 1)(x - 7)(x - 1).$$

In this example the cubic polynomial has been factorised into a product of three linear factors.

Method notes

Follow the same method as the previous example.

Stop and think 1

Explain why writing the polynomial from the previous example in factorised form is helpful in sketching the graph of $y = x^3 - 7x^2 - x + 7$

Factor theorem

In the previous section we found that $(x + 5)$ is a factor of

$$f(x) = 3x^3 + 14x^2 - 3x + 10$$

Therefore we can write

$$f(x) = (x + 5)(3x^2 - x + 2).$$

If $(x + 5) = 0$ we can say that

$$f(x) = (0)(3x^2 - x + 2) = 0 \qquad (1)$$

But $x + 5 = 0$ means that $x = -5$ (2)

Substituting $x = -5$ from equation (2) into equation (1) we get $f(-5) = 0$

Substituting for $x = -5$ in the original cubic polynomial gives

$$f(-5) = 3(-5)^3 + 14(-5)^2 - 3(-5) + 10 = 0$$

Similarly we showed that $(x + 1)$ is a factor of the polynomial $g(x) = x^3 - 7x^2 - x + 7$ so we can write

$$g(x) = (x + 1)(x^2 - 8x + 7)$$

If $(x + 1) = 0$ we know that

$$g(x) = (0)(x^2 - 8x + 7) = 0 \qquad (3)$$

But $x + 1 = 0$ means that $x = -1$ (4)

Substituting $x = -1$ from equation (4) into equation(3) we get $g(-1) = 0$

Substituting $x = -1$ in the original cubic polynomial gives

$$g(-1) = (-1)^3 - 7(-1)^2 - (-1) + 7 = 0$$

These examples illustrate the **factor theorem** for polynomials which states that: if$(x - p)$ is a factor of $f(x)$ where $f(x)$ is a polynomial then $f(p) = 0$

Essential notes

If there is no remainder when one polynomial is divided by another then this means the dividing polynomial is a factor of the other polynomial.

Essential notes

If $(x + 5)$ is a factor as in the example on page 7, we can compare with $(x - p)$ to get $p = -5$

Example

Use the factor theorem to show that

a) $(x + 3)$ is a factor of $f(x) = 5x^3 + 16x^2 + 7x + 12$

b) $(x - 4)$ is a factor of $g(x) = x^4 - 3x^3 - 3x^2 - x - 12$

Answer

a) $f(x) = 5x^3 + 16x^2 + 7x + 12$

$$f(-3) = 5(-3)^3 + 16(-3)^2 + 7(-3) + 12$$
$$= -135 + 144 - 21 + 12 = 0$$

Therefore $(x + 3)$ is a factor of $5x^3 + 16x^2 + 7x + 12$

b) $g(x) = x^4 - 3x^3 - 3x^2 - x - 12$

$$g(4) = (4)^4 - 3(4)^3 - 3(4)^2 - (4) - 12$$
$$= 256 - 192 - 48 - 4 - 12 = 0$$

Therefore $(x - 4)$ is a factor of $x^4 - 3x^3 - 3x^2 - x - 12$

Method notes

a) To test whether $(x + 3)$ is a factor then $x + 3 = 0$ so $x = -3$

We must now evaluate $f(-3)$. If $f(-3) = 0$ then $(x + 3)$ is a factor.

b) To test whether $(x - 4)$ is a factor then $x - 4 = 0$ so $x = 4$

If $g(4) = 0$ then $(x - 4)$ is a factor.

Example

Given that $(x+1)$ and $(x-2)$ are factors of $f(x)=2x^3+px^2+qx-10$ find the values of p and q.

Answer

Step 1: Using the factor theorem if $(x+1)$ is a factor we can say that $f(-1)=2(-1)^3+p(-1)^2+q(-1)-10=0$

Step 2: Simplify the equation to give $p-q=12$ (1)

Step 3: Using the factor theorem if $(x-2)$ is a factor we can say that $f(2)=2(2)^3+p(2)^2+q(2)-10=0$

Step 4: Simplify the equation to give $4p+2q=-6$ (2)

Step 5: Solve the simultaneous equations (1) and (2)

(1) $\times$ 2 gives $2p-2q=24$

Adding to (2) $4p+2q=-6$

Gives $6p=18$ so $p=3$

Step 6: Substituting $p=3$ into equation (1) gives $3-q=12$ so $q=-9$

Exam tips

You should learn how to use the factor theorem as you are likely to be tested on this in the examination. This example and the next illustrate two different types of examination questions.

Example

Show that $(x-1)$ is NOT a factor of $g(x)=4x^3-3x^2-3$

Answer

Step 1: If $x-1$is a factor then $x-1=0$ so $x=1$ and $g(1)$ must be equal to 0

Step 2: If $g(x)=4x^3-3x^2-3$ we know that $g(1)=4(1)^3-3(1)^2-3=-2$

Step 3: State the answer: as $g(1)\neq 0$ then $(x-1)$ is not a factor of $g(x)=4x^3-3x^2-3$

Remainder theorem

In the previous example we showed that $(x-1)$ is **not** a factor of $g(x)=4x^3-3x^2-3$ because $g(1)=-2\neq 0$

The following example shows that there is a connection between the remainder and the dividing polynomial.

Example

Find the remainder when $g(x)=4x^3-3x^2-3$ is divided by $(x-1)$.

Answer

$$
\begin{array}{r l}
 & 4x^2 + x + 1 \quad \leftarrow \boxed{\text{quotient}} \\
(x-1) \,\big) & 4x^3 - 3x^2 - 3 \\
 & \underline{4x^3 - 4x^2} \\
 & x^2 - 3 \\
 & \underline{x^2 - x} \\
 & x - 3 \\
 & \underline{x - 1} \\
 & -2 \quad \leftarrow \boxed{\text{remainder} = -2}
\end{array}
$$

Method notes

Follow the method explained in the previous examples for algebraic division.

On page 10 we saw that $(x - 1)$ was not a factor of $g(x)$ because $g(1) \neq 0$

In the previous example we saw that when $g(x) = 4x^3 - 3x^2 - 3$ is divided by $(x - 1)$ the remainder is -2

If we evaluate $g(1) = 4(1)^3 - 3(1)^2 - 3 = -2$

This is an example of the **remainder theorem** which states that if a polynomial $g(x)$ is divided by $(ax - b)$ then the remainder is $g\left(\frac{b}{a}\right)$.

Essential notes

In this example the divisor is $(x - 1)$. If we compare $(x - 1)$ with $(ax - b)$ then $a = 1$ and $b = 1$

In this case $g\left(\frac{b}{a}\right) = g(1) = -2$ which is the remainder.

Example

Find the remainder when

a) $3x^3 - 5x^2 + 4x - 1$ is divided by $(x - 2)$

b) $8x^4 - 8x^3 + 4x^2 - 5x + 3$ is divided by $(2x + 1)$.

Answer

a) **Step 1**: From the remainder theorem statement:

$f(x) = 3x^3 - 5x^2 + 4x - 1$

and if this is divided by $(x - 2)$ then $a = 1$ and $b = 2$

Step 2: Solve $x - 2 = 0$ to give $x = 2$

Step 3: Using the remainder theorem the remainder is $f\left(\frac{b}{a}\right)$

In this case $f\left(\frac{b}{a}\right) = f\left(\frac{2}{1}\right) = f(2)$

Step 4: Evaluate $f(2) = 3(2)^3 - 5(2)^2 + 4(2) - 1 = 24 - 20 + 8 - 1 = 11$

Step 5: State the answer: the remainder when $3x^3 - 5x^2 + 4x - 1$ is divided by $(x - 2)$ is 11

b) $g(x) = 8x^4 - 8x^3 + 4x^2 - 5x + 3$

$$g\left(-\frac{1}{2}\right) = 8\left(-\frac{1}{2}\right)^4 - 8\left(-\frac{1}{2}\right)^3 + 4\left(-\frac{1}{2}\right)^2 - 5\left(-\frac{1}{2}\right) + 3$$

$$= 8\left(\frac{1}{16}\right) - 8\left(-\frac{1}{8}\right) + 4\left(\frac{1}{4}\right) - 5\left(-\frac{1}{2}\right) + 3$$

$$= 8$$

The remainder when $8x^4 - 8x^3 + 4x^2 - 5x + 3$ is divided by $(2x + 1)$ is 8

Exam tips

You must learn how to use the remainder theorem as you are likely to be tested on this in the Core 2 examination.

Method notes

In (b) use the same method as in (a).

Solve $2x + 1 = 0$ to give $x = -\frac{1}{2}$

Evaluate $g\left(-\frac{1}{2}\right)$ to find the remainder.

Factorising cubic polynomials

When you studied quadratics in Core 1 you learnt that a quadratic has two linear factors, one repeated linear factor or no factors depending on the value of the **discriminant** $b^2 - 4ac$. These factors are associated with the number of roots of the quadratic.

Fig. 1.1
The roots of a quadratic equation.

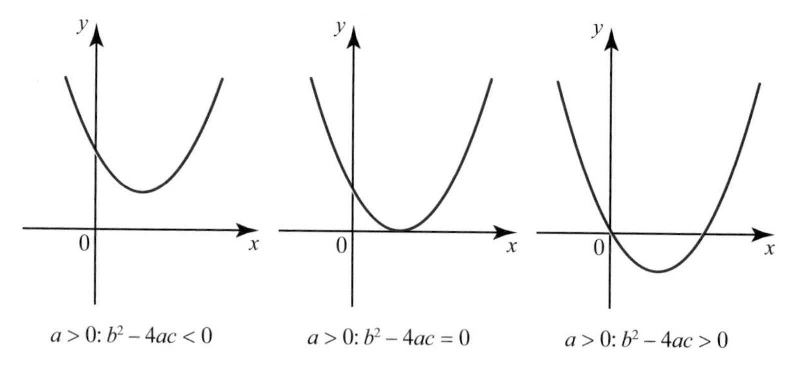

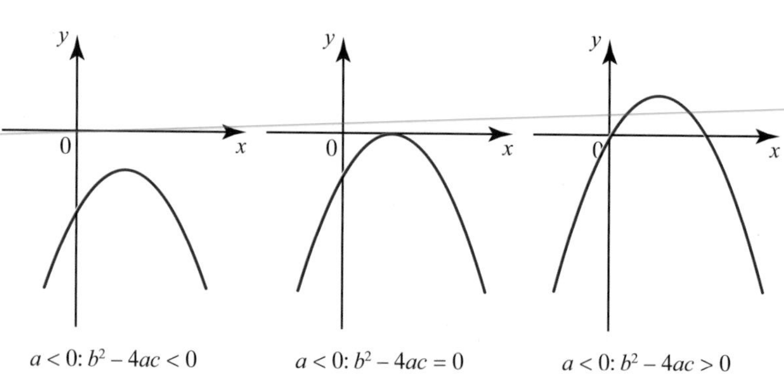

Essential notes

The roots are the values of x where the quadratic cuts or touches the x-axis. The definition and use of the discriminant was covered in Core 1.

The following sketches of different cubic functions show the relationship between the number of roots of the cubic equation and where the graph crosses the x-axis.

For cubic polynomials there is always one root and hence one linear factor.

Essential notes

Sketching functions was covered in Core 1.

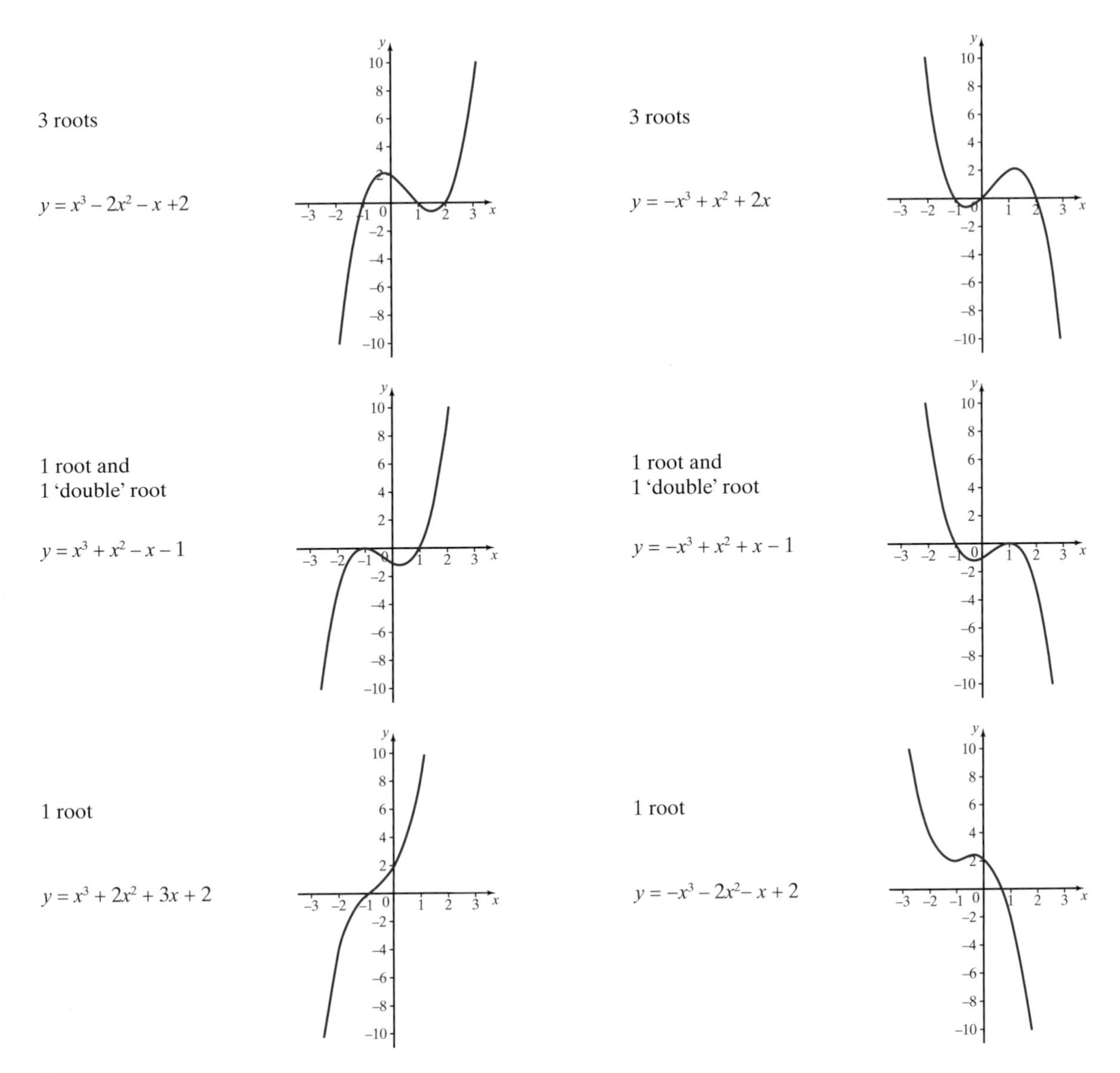

Fig. 1.2
The roots of a cubic equation.

A cubic can always be factorised as the product of a linear factor and a quadratic factor. Whether the cubic has three linear factors, two linear factors or one linear factor then depends on how many roots the quadratic factor has.

Stop and think 2

Explain why there is always one root and hence one linear factor of a cubic polynomial.

Example

Show that $(x - 1)$ is a linear factor of the following cubic polynomials. Hence factorise the cubic polynomials into three linear factors, two linear factors or one linear factor.

a) $x^3 - 5x^2 + 8x - 4$

b) $x^3 - 1$

c) $x^3 - 6x^2 + 11x - 6$

Answer

a) $p(x) = x^3 - 5x^2 + 8x - 4$

$p(1) = (1)^3 - 5(1)^2 + 8(1) - 4 = 1 - 5 + 8 - 4 = 0$

$(x - 1)$ is a factor of $x^3 - 5x^2 + 8x - 4$

$$
\begin{array}{r l}
 & x^2 - 4x + 4 \\
(x-1) & \overline{)x^3 - 5x^2 + 8x - 4} \\
 & \underline{x^3 - x^2} \\
 & -4x^2 + 8x - 4 \\
 & \underline{-4x^2 + 4x} \\
 & 4x - 4 \\
 & \underline{4x - 4} \\
 & 0
\end{array}
$$

$x^3 - 5x^2 + 8x - 4 = (x - 1)(x^2 - 4x + 4)$

$x^2 - 4x + 4 = (x - 2)(x - 2)$

Therefore $x^3 - 5x^2 + 8x - 4 = (x - 1)(x - 2)(x - 2)$

The cubic polynomial $p(x) = x^3 - 5x^2 + 8x - 4$ has three linear factors, one of the factors $(x - 2)$ occurs twice and is called a **repeated factor**.

Method notes

$p(x)$ is the usual function notation here instead of $f(x)$.

If $(x - 1)$ is a factor then using the factor theorem $p(1) = 0$.

Algebraic division of $p(x)$ by $(x - 1)$ then gives the quadratic factor.

Factorisation of the quadratic factor gives two linear factors.

The cubic polynomial can then be written as the product of three linear factors one of which is a 'repeated' factor in this case.

b) $p(x) = x^3 - 1$

$p(1) = (1)^3 - 1 = 1 - 1 = 0$

$(x - 1)$ is a factor of $x^3 - 1$

$$
\begin{array}{r}
x^2 + x + 1 \\
(x-1)\,\overline{)\,x^3 \qquad\quad - 1} \\
\underline{x^3 - x^2} \qquad\quad \\
x^2 \quad - 1 \\
\underline{x^2 - x} \quad\;\; \\
x - 1 \\
\underline{x - 1} \\
0
\end{array}
$$

$x^3 - 1 = (x - 1)(x^2 + x + 1)$

$(x^2 + x + 1)$ does not have any factors because $b^2 - 4ac = 1 - 4 = -3 < 0$
The cubic polynomial $p(x) = x^3 - 1$ has one linear factor.

c) $p(x) = x^3 - 6x^2 + 11x - 6$

$p(1) = (1)^3 - 6(1)^2 + 11(1) - 6 = 1 - 6 + 11 - 6 = 0$

$(x - 1)$ is a factor of $x^3 - 6x^2 + 11x - 6$

$$
\begin{array}{r}
x^2 - 5x + 6 \\
(x-1)\,\overline{)\,x^3 - 6x^2 + 11x - 6} \\
\underline{x^3 - x^2} \qquad\qquad\quad \\
-5x^2 + 11x - 6 \\
\underline{-5x^2 + 5x} \quad\;\; \\
6x - 6 \\
\underline{6x - 6} \\
0
\end{array}
$$

$x^3 - 6x^2 + 11x - 6 = (x - 1)(x^2 - 5x + 6)$

$x^2 - 5x + 6 = (x - 2)(x - 3)$

Therefore $x^3 - 6x^2 + 11x - 6 = (x - 1)(x - 2)(x - 3)$

The cubic polynomial $p(x) = x^3 - 6x^2 + 11x - 6$ has three distinct linear factors.

Method notes

Given $p(x) = x^3 - 1$ we can see by inspection that if $x = 1$ then $p(x) = 1^3 - 1 = 0$

Using the factor theorem then gives $(x - 1)$ as a factor.

Algebraic division of $p(x)$ by $(x - 1)$ will then give the quadratic factor.

In this case the quadratic factor cannot be factorised because the discriminant < 0

We conclude that the cubic polynomial has only one linear factor.

Method notes

If $(x - 1)$ is a factor then using the factor theorem $p(x) = 0$

Algebraic division of $p(x)$ by $(x - 1)$ then gives the quadratic factor.

Factorisation of the quadratic factor gives two linear factors.

We conclude that the cubic polynomial has three distinct (or different) linear factors.

Stop and think 3

Using the example above, write down the solutions of the following cubic equations where x is a real number.

a) $x^3 - 5x^2 + 8x - 4$

b) $x^3 - 1$

c) $x^3 - 6x^2 + 11x - 6$

Stop and think answers

1. $x^3 - 7x^2 - x + 7 = (x + 1)(x - 7)(x - 1)$.

   The graph $y = x^3 - 7x^2 - x + 7$ crosses the x-axis when $y = 0$ and $x = -1, 7$, and 1

   So the 3 points on the curve are $(-1, 0)(1, 0)$ and $(7, 0)$

2. Any cubic equation $f(x) = 0$ can be factorised as the product of a linear factor and a quadratic factor. This means that the linear factor will always give a solution to the equation $f(x) = 0$ so there is always one root. Any further roots of the equation $f(x) = 0$ will be determined by the solutions of the quadratic factor.

3. a) To solve $x^3 - 5x^2 + 8x - 4 = 0$ factorise $x^3 - 5x^2 + 8x - 4 = (x - 1)(x - 2)(x - 2)$

   If we solve $(x - 1)(x - 2)(x - 2) = 0$ the solutions are: $x = 1$, $x = 2$ twice.

   b) To solve $x^3 - 1 = 0$ factorise $x^3 - 1 = (x - 1)(x^2 + x + 1)$

   we solve $(x - 1)(x^2 + x + 1) = 0$ to give $x - 1 = 0$ or $(x^2 + x + 1) = 0$

   $x = 1$ is a real root but in $(x^2 + x + 1) = 0$ the discriminant is negative so there are no further real roots.

   c) To solve $x^3 - 6x^2 + 11x - 6 = 0$ factorise $x^3 - 6x^2 + 11x - 6 = (x - 1)(x - 2)(x - 3)$

   If we solve $(x - 1)(x - 2)(x - 3) = 0$ the solutions are: $x = 1$, $x = 2$, $x = 3$.

Review of the geometry of straight lines

There are several properties of straight lines that are used when investigating the geometry of circles. You have met these before in GCSE and in Core 1.

In Figure 2.1, A and B are two points on a straight line. The coordinates of A are (x_1, y_1) and the coordinates of B are (x_2, y_2). P is the midpoint of the line AB.

Fig. 2.1
P is the midpoint of the line AB.

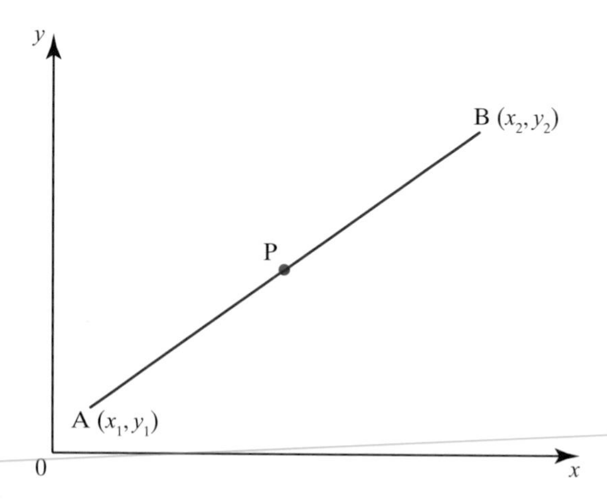

Property 1
The length of the line AB is $\sqrt{(x_2 - x_1)^2 + (y_2 - y_1)^2}$

Property 2
The coordinates of the midpoint P are $\left(\frac{x_2 + x_1}{2}, \frac{y_2 + y_1}{2}\right)$

Property 3
The gradient (or slope) of the line AB is $\frac{y_2 - y_1}{x_2 - x_1}$

Property 4
If a straight line L has gradient m then the slope of a straight line that is perpendicular to L is $-\frac{1}{m}$.

Example
The line AB is a diameter of a circle, where A and B are (5, 7) and (–1, 3) respectively.

a) Find the coordinates of the centre of the circle.

b) Find the length of the diameter of the circle and hence write down the length of the radius.

c) Find the gradient of the diameter of the circle.

d) Find the equation of the line that passes through A and B.

e) Find the equation of the line perpendicular to AB that passes through the centre of the circle.

Answer

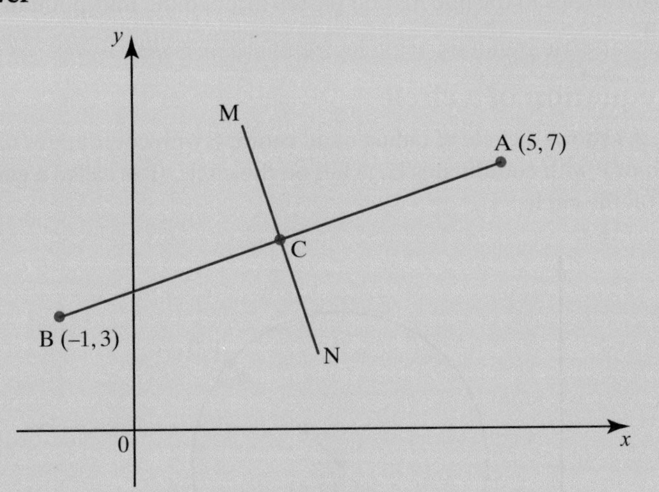

Fig. 2.2

a) The centre of the circle, C, is at the mid-point of AB.

Step 1: Point A has coordinates (5, 7) and B has coordinates (–1, 3) so $x_1 = 5, y_1 = 7, x_2 = -1, y_2 = 3$

Step 2: Using property 2 we can say that the coordinates of the mid-point C are $\left(\frac{5 + (-1)}{2}, \frac{7 + 3}{2}\right) = (2, 5)$

b) **Step 1**: Using property 1 we can say that the length of the diameter AB is $\sqrt{(5 - (-1))^2 + (7 - 3)^2} = \sqrt{6^2 + 4^2}$

$= \sqrt{52}$

$= 2\sqrt{13}$

Step 2: Length of radius is half the length of diameter $= \sqrt{13}$

c) **Step 1**: Let the gradient of the diameter of the circle be m

Step 2: Using property 3 we can say that $m = \frac{7 - 3}{5 - (-1)} = \frac{4}{6} = \frac{2}{3}$

d) The equation of the line through AB is

$$\frac{y - 7}{x - 5} = \frac{2}{3}$$

$\Rightarrow 3(y - 7) = 2(x - 5)$

$\Rightarrow \quad y = \frac{2}{3}x + \frac{11}{3}$

e) If the line MN through C is perpendicular to AB using property 4 we can say that the slope of MN is $-\frac{1}{m} = -\frac{3}{2}$

The equation of the line through MN is therefore $\frac{y - 5}{x - 2} = -\frac{3}{2}$

which means $2(y - 5) = -3(x - 2)$ which means $y = -\frac{3}{2}x + 8$

Method notes

Always draw a clear diagram.

Find the centre of the circle which is the mid-point of the diameter (AB) and label it C.

Draw a line perpendicular to the diameter through C and label it (here it is MN).

a) Property 2 of straight line geometry states that mid-point coordinates are: $\left(\frac{x_2 + x_1}{2}, \frac{y_2 + y_1}{2}\right)$

b) Property 1 of straight line geometry states that the length of the line AB is $\sqrt{(x_2 - x_1)^2 + (y_2 - y_1)^2}$

Using rules of indices $\sqrt{52} = \sqrt{4} \times \sqrt{13} = 2\sqrt{13}$ which gives the answer as a surd.

c) Property 3 of straight line geometry states that the gradient (or slope) of the line AB is $\frac{y_2 - y_1}{x_2 - x_1}$

d) Equation of a line with gradient m passing through the point (x_1, y_1) is $y - y_1 = m(x - x_1)$

e) Property 4 of straight line geometry states that the product of the gradients of perpendicular lines $= -1$.

The line MN is called the **perpendicular bisector** of the line AB since it is at right angles to the line AB and passes through the mid-point of the line AB.

The equation of a circle

Figure 2.3 shows a circle of radius r and centre C with coordinates (a, b). The point P with coordinates (x, y) lies on the circle. (P is called a **general point** on the circle.)

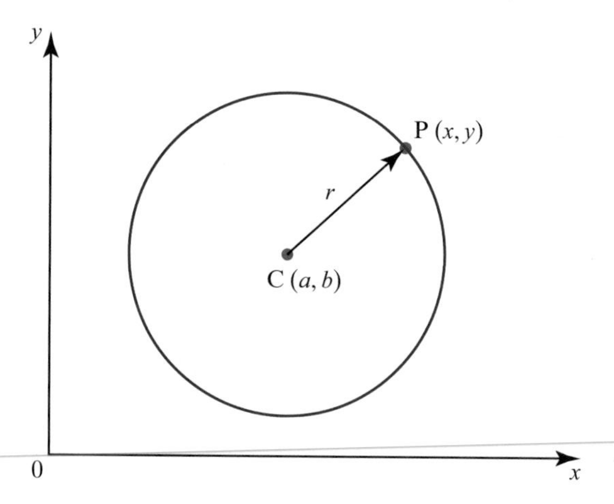

Fig. 2.3
A circle radius r, centre C, showing a general point on the circle.

The length of the line CP is $\sqrt{(x-a)^2+(y-b)^2}=r$ so $(x-a)^2+(y-b)^2=r^2$

This is the general form of the **equation of a circle** with centre (a, b) and radius r.

Method notes

Use Property 1 for the geometry of straight lines with

$x_2=x \quad y_2=y \quad x_1=a \quad y_1=b$

Exam tips

This formula is in your formula booklet. You need to learn how to apply it.

Example

Write down the equation of the following circles:

a) centre (3, 2), radius 3

b) centre (5, −6), radius 5

c) centre $(2a, 7a)$, radius $5a$

Answer

a) $(x-3)^2+(y-2)^2=3^2=9$

b) $(x-5)^2+(y-(-6))^2=5^2=25$

so $(x-5)^2+(y+6)^2=25$

c) $(x-2a)^2+(y-7a)^2=(5a)^2=25a^2$

Method notes

List the information given in terms of the general form of the equation of a circle.

a) $a=3, b=2, r=3$
b) $a=5, b=-6, r=5$
c) $a=2a, b=7a, r=5a$

The following example illustrates that, if given the equation of a circle, you can find the coordinates of the centre and the radius.

Example

a) Write down the coordinates of the centre and radius of the circle with equation:

$(x + 7)^2 + (y - 2)^2 = 65$

b) Show that the point (0, −2) lies on the circle.

Answer

a) Comparing the general equation of a circle $(x - a)^2 + (y - b)^2 = r^2$ with $(x + 7)^2 + (y - 2)^2 = 65$ the centre of the circle has coordinates (−7, 2) and $r = \sqrt{65}$

b) (0, −2) lies on the circle since $(0 + 7)^2 + (-2 - 2)^2 = 49 + 16 = 65$

Method notes

a) $-a = 7$ and $-b = -2$ so $a = -7$ and $b = 2$
The centre has coordinates (a, b)
$r^2 = 65$ so $r = \sqrt{65}$

b) If the point (0, −2) lies on the circle the coordinates (0, −2) must satisfy the equation of the circle.

Example

The line AB is a diameter of a circle, where A and B are points with coordinates (2, 5) and (−6, −1) respectively.

Find the equation of the circle.

Answer

Step 1: The general formula for the equation of a circle is $(x - a)^2 + (y - b)^2 = r^2$ where (a, b) is the centre and r is the radius so we need to find a, b, and r.

Step 2: Using property 1 of straight line geometry, the length of the diameter AB is

$\sqrt{((-6) - 2)^2 + ((-1) - 5)^2} = \sqrt{(64 + 36} = \sqrt{100} = 10$

Step 3: The radius is half the length of the diameter therefore $r = 5$

Step 4: The centre of a circle is the mid-point of the diameter so using property 2 of straight line geometry the centre of the circle is:

$\left(\frac{-6 + 2}{2}, \frac{-1 + 5}{2}\right) = (-2, 2)$ so $a = -2$, $b = 2$

Step 5: Substituting the values for a, b and r into the general equation from step 1 means the equation of the circle is

$(x - (-2))^2 + (y - 2)^2 = 5^2$ so $(x + 2)^2 + (y - 2)^2 = 25$

Essential notes

'Respectively' means that point A has coordinates (2, 5) that is, $x_1 = 2$ $y_1 = 5$ and the point B has coordinates (−6, 1).

Exam tips

You will need to recognise all three forms of the equation of the circle. You must also learn which form to use, depending on the information given in the examination question.

Different forms of the equation of a circle

The general form of the equation of the circle was stated in the previous section as $(x-a)^2+(y-b)^2=r^2$ where the circle has centre (a, b) and radius r.

Two other forms of the equation can be developed from this general form.

If $a=0$ and $b=0$ this means the centre of the circle would be the origin (0, 0). Substituting these values into the equation above gives a **second form** of the equation:

$$x^2+y^2=r^2$$

where the circle has centre (0, 0) and radius r.

If the general equation $(x-a)^2+(y-b)^2=r^2$ is expanded it becomes

$$x^2+y^2-2ax-2by+(a^2+b^2-r^2)=0$$

But a, b, and r are all constants so $(a^2+b^2-r^2)$ is a constant.

If we let $(a^2+b^2-r^2)=c$ then we have a **third form** of the equation of the circle:

$$x^2+y^2-2ax-2by+c=0$$

where the circle has centre (a, b), $c=(a^2+b^2-r^2)$ so $r^2=a^2+b^2-c$.

Example

Find the coordinates of the centre and the radius of the circle with equation $x^2+6x+y^2+8y-11=0$

Answer

Step 1: Decide which form of the equation of the circle to use. The equation in the question compares most easily with the form $x^2+y^2-2ax-2by+c=0$ where (a, b) is the centre and $r^2=a^2+b^2-c$.

Step 2: Compare the two forms $x^2+6x+y^2+8y-11=0$ and $x^2+y^2-2ax-2by+c=0$:

$-2a=6$ so $a=-3$

$-2b=8$ so $b=-4$ and $c=-11$

so centre of the circle is $(-3, -4)$

Step 3: Substitute values for a, b and c from step 2 into $r^2=a^2+b^2-c$ from step 1 to give

$r^2=(-3)^2+(-4)^2-(-11)$

$=9+16+11=36$ so $r=6$ so radius of the circle is 6

Stop and think 1

Find the coordinates of the centre and the radius of the circle with equation $x^2 + 6x + y^2 + 8y - 11 = 0$ by using $(x - a)^2 + (y - b)^2 = r^2$ as the equation of a circle.

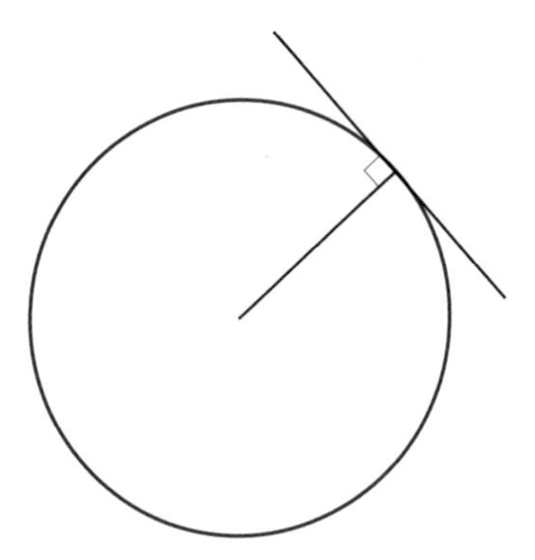

Fig. 2.4
A circle and tangent showing the 90° angle.

The geometry of lines and circles

You will be familiar with several circle theorems from GCSE mathematics. The algebra of circles allows us to explore these theorems in a different way.

Tangents and circles

A **tangent** to a circle is a straight line which touches the circle at one point only. The angle between a tangent and a radius is 90° as shown in Figure 2.4.

Essential notes

We say that the radius and tangent are perpendicular.

Solution of quadratic equations using the discriminant $b^2 - 4ac$ and the solution of quadratic inequalities was covered in Core 1.

Example

Find the values of c for which the line $y = 2x + c$

a) touches the circle with equation $x^2 + y^2 - 2x = 0$

b) intersects the circle at two points

c) does not meet the circle.

Answer

Where a line and curve intersect the two equations are both true simultaneously.

Substitute for $y = 2x + c$ into $x^2 + y^2 - 2x = 0$ to give $x^2 + (2x + c)^2 - 2x = 0$

so $x^2 + 4x^2 + 4xc + c^2 - 2x = 0$

so $5x^2 + 2(2c - 1)x + c^2 = 0$

The number of solutions of this quadratic equation depends on which part of the question you are answering.

a) The circle and line touch once if the quadratic equation has one solution, that is, a repeated root so if $b^2 - 4ac = 0$:

$$\Rightarrow 4(2c - 1)^2 - 20c^2 = 0$$
$$\Rightarrow 4(4c^2 - 4c + 1) - 20c^2 = 0$$
$$\Rightarrow -4c^2 - 16c + 4 = 0$$
$$\Rightarrow 4c^2 + 16c - 4 = 0$$
$$\Rightarrow c^2 + 4c - 1 = 0$$
$$\Rightarrow c = -2 + \sqrt{5} \text{ or } c = -2 - \sqrt{5}$$

Method notes

a) If the line **touches** the circle there is **one** common point on the line and the circle.

Continued on the next pages

Method notes

b) Distinct means different.

c) If the line does not meet the circle there are no common points.

b) The circle and line intersect at two points if the quadratic has two distinct roots so if $b^2 - 4ac > 0$:

$\Rightarrow -4c^2 - 16c + 4 > 0$

$\Rightarrow 4c^2 + 16c - 4 < 0$

$\Rightarrow c^2 + 4c - 1 < 0$

$\Rightarrow -2 - \sqrt{5} < c < -2 + \sqrt{5}$

c) The line and circle do not meet if the quadratic has no solutions so if $b^2 - 4ac < 0$:

$\Rightarrow (2(2c - 1))^2 - 4(5)\,c^2 < 0$

$\Rightarrow 16c^2 + 4 - 16c - 20c^2 < 0$

$\Rightarrow 4 - 16c - 4c^2 < 0$

$\Rightarrow c < -2 - \sqrt{5}$ or $c > -2 + \sqrt{5}$

You need to be able to solve problems by using information about intersecting lines and circles in a different way. The following example illustrates this.

Method notes

You can solve the simultaneous equations by either substituting for x or substituting for y.

Example

The line $x + 3y - 11 = 0$ touches the circle $(x + 1)^2 + (y + 6)^2 = 90$ at the point P.

a) Find the coordinates of the point P.

b) Show that the radius through P is perpendicular to the tangent at P.

Answer

The line **touches** the circle so as shown in the last example there will be **one repeated root** of the quadratic equation formed by using the equation of the line and the equation of the circle simultaneously.

a) **Step 1**: From the equation of the line $x + 3y - 11 = 0$ we can say that $x = 11 - 3y$

Step 2: Substituting the value of x from step 1 into the equation of the circle gives $(11 - 3y + 1)^2 + (y + 6)^2 = 90$

Step 3: Simplifying the algebra gives $(12 - 3y)^2 + (y + 6)^2 = 90$

$\Rightarrow 144 - 72y + 9y^2 + y^2 + 12y + 36 = 90$

$\Rightarrow 10y^2 - 60y + 90 = 0$

$\Rightarrow y^2 - 6y + 9 = 0$

Step 4: Factorising the quadratic in step 3 gives $(y - 3)^2 = 0$ so $y = 3$ (twice) which represents a repeated root .

Step 5: Substitute $y = 3$ from step 4 into the equation of the line gives $x = 11 - 3y$ and $x = 2$

Therefore the coordinates of the point P where the line and circle touch are (2, 3).

b) **Step 1**: Using the form of the equation of a circle $(x-a)^2+(y-b)^2=r^2$ and comparing this with $(x+1)^2+(y+6)^2=90$ gives $a=-1$ and $b=-6$ and so the centre C is the point $(-1, -6)$

Step 2: Any radius must pass through the centre of the circle hence this radius must pass through P(2, 3) and C(−1, −6).

Step 3: Using property 3 of the geometry of straight lines so the gradient of the radius $CP=\frac{-6-3}{-1-2}=\frac{-9}{-3}=3$

Step 4: The line $x+3y-11=0$ touches the circle at the point P as shown in a) hence this line is a tangent to the circle at the point P.

Step 5: Rearranging $x+3y-11=0$ to the form $y=mx+c$ where m is the gradient means the gradient of the tangent at P is $-\frac{1}{3}$

Step 6: Using the values of the gradients in step 3 and step 5 shows that the product of these gradients is −1 and therefore the radius through P is perpendicular to the tangent at P.

Chords and circles

A **chord** is a line which joins any two points on the circumference of a circle. A chord divides a circle into two segments. The perpendicular line through the centre of the circle, C, to a chord, AB, bisects the chord as shown in Figure 2.5 below.

CM is called the perpendicular bisector of AB.

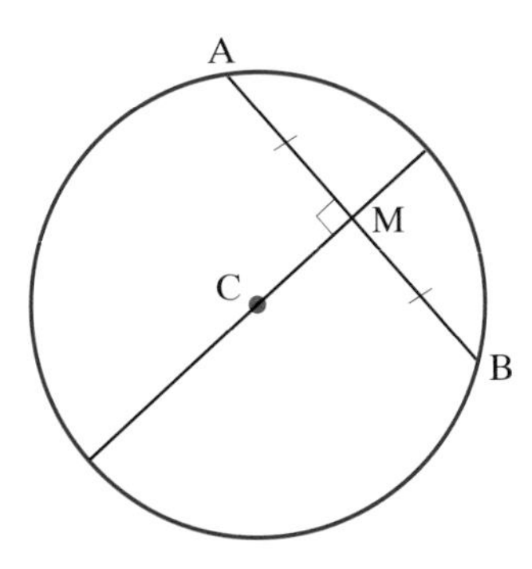

Fig. 2.5
M bisects the chord AB or M is the mid-point of AB.

Example

The points A and B with coordinates (–4, 3) and (7, 4) lie on the circle with equation

$$(x-2)^2 + (y+2)^2 = 61.$$

a) Find the coordinates of the mid-point, M, of the chord AB.

b) Find the slope of the chord AB.

c) Show that the diameter through the point M is perpendicular to the chord and hence show that the perpendicular line through the centre of the circle to a chord, AB, bisects the chord.

Answer

Fig. 2.6

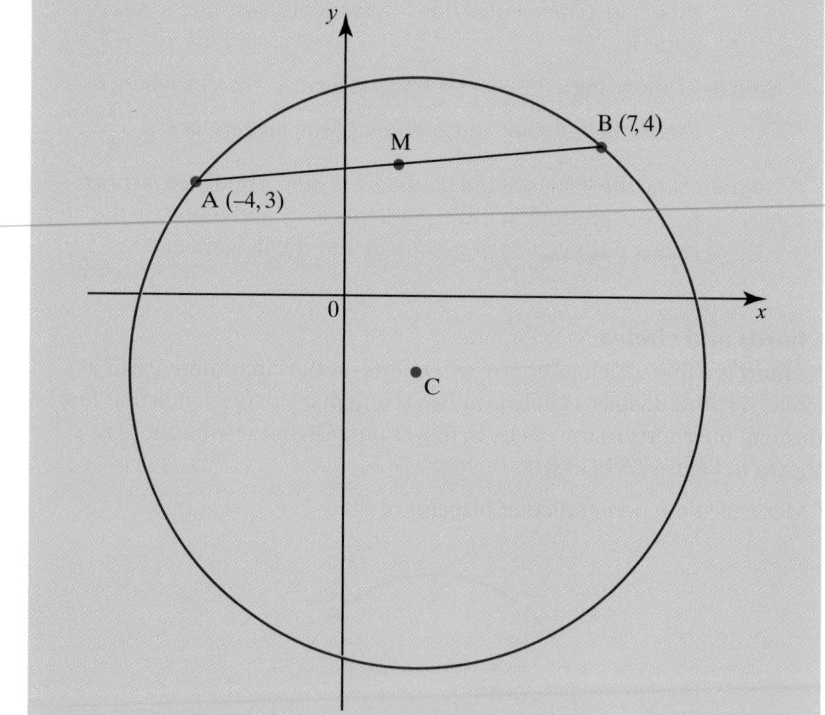

Method notes

Draw a diagram from the information given in the question as shown in Figure 2.6.

a) The coordinates of the mid-point of AB are

$$\left(\frac{7-4}{2}, \frac{4+3}{2}\right) = \left(\frac{3}{2}, \frac{7}{2}\right) = (1.5, 3.5)$$

b) The gradient of the chord AB is: $\dfrac{4-3}{7-(-4)} = \dfrac{1}{11}$

c) From the equation of the circle, $(x-2)^2 + (y+2)^2 = 61$ and the coordinates of the centre C are (2, –2).

The gradient of the diameter through M and C is:

$$\frac{\frac{7}{2}-(-2)}{\frac{3}{2}-2} = \frac{\frac{11}{2}}{-\frac{1}{2}} = -11$$

Hence the diameter through M and C is perpendicular to the chord AB because the product of their gradients is –1 and the diameter bisects the chord because M is its midpoint.

Angle in a semi-circle

The angle subtended (formed) by a diameter at a point on the circumference is 90°. This angle is often called the **angle in a semi-circle** and is illustrated in Figure 2.7.

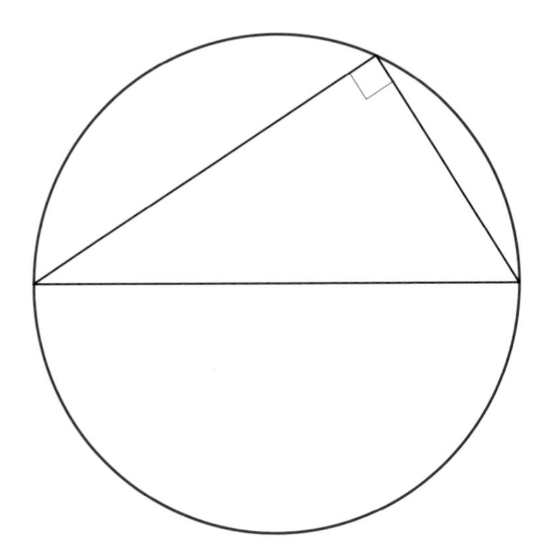

Fig. 2.7
Angle in a semi-circle is 90°.

Example

The points A (−1, 8) and B (5, 4) form the diameter of a circle.

a) Find the equation of the circle.

b) Show that the point D (4, 9) lies on the circle.

c) Show that the chords AD and BD are perpendicular.

Answer

a) The centre of the circle is at the mid-point of the diameter AB

$$\left(\frac{-1+5}{2}, \frac{8+4}{2}\right) = (2, 6)$$

The centre of the circle has coordinates (2, 6).

The diameter of the circle is the length of AB.

$$\sqrt{(5-(-1))^2+(4-8)^2} = \sqrt{36+16} = \sqrt{52} = 2\sqrt{13}$$

The radius of the circle is $\sqrt{13}$

The equation of the circle is $(x-2)^2+(y-6)^2 = 13$

Method notes

Draw a diagram of the information given as shown in Figure 2.8 below then decide from this information which of the forms of the equation of a circle to use.

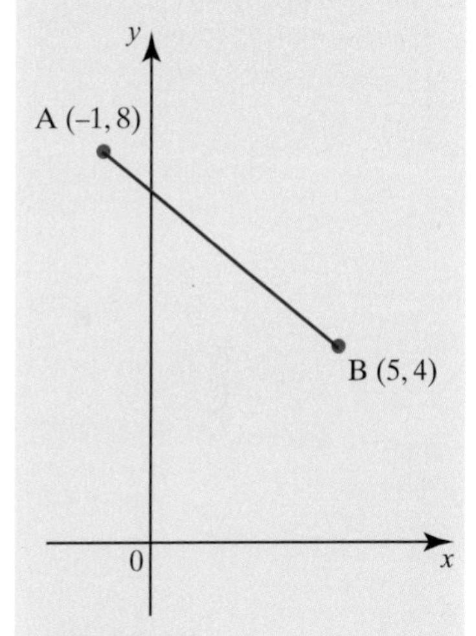

Fig. 2.8

a) Use property 2 to find the mid-point of the diameter. This point is the centre of the circle.

Use property 1 to find the length of the diameter AB and the radius is half the length of AB.

Use the first form of the equation of the circle.

Method notes

c) Use property 4 to show that lines are perpendicular (or at 90°) to each other.

b) If a point lies on the circle in (a) its coordinates must satisfy the equation of that circle.

Satisfy means that the coordinates of the given point must make the equation of the circle mathematically true.

D is the point (4, 9) so when $x = 4$ and $y = 9$ are substituted into the equation of the circle this must be a correct mathematical result.

The equation of the circle is $(x - 2)^2 + (y - 6)^2 = 13$

Replacing x by 4 and y by 9 gives $(4 - 2)^2 + (9 - 6)^2 = 2^2 + 3^2 = 4 + 9 = 13$

This is the same result as the equation of the circle so the point D lies on the circle.

c) A is the point (−1, 8), D is the point (4, 9) therefore the gradient of the chord AD $= \frac{9 - 8}{4 - (-1)} = \frac{1}{5}$

B is the point (5, 4) therefore the gradient of the chord BD $= \frac{9 - 4}{4 - 5} = \frac{5}{-1} = -5$

Since the gradient of AD × the gradient of BD = −1 the chords AD and BD are perpendicular. So the diameter AB subtends a right angle (90°) at the point D on the circumference.

Stop and think answers

1. *To find the coordinates of the centre and radius of the circle $x^2 + 6x + y^2 + 8y - 11 = 0$ use the method of completing the square as covered in Core 1 so $(x + 3)^2 - 9 + (y + 4)^2 - 16 - 11 = 0$*

   $(x + 3)^2 + (y + 4)^2 = 11 + 16 + 9$ so $(x + 3)^2 + (y + 4)^2 = 36$

   comparing with $(x + a)^2 + (y + b)^2 = r^2$

   so centre is (−3, −4) and radius is 6

In Core 1 you studied the properties of arithmetic sequences and series. An arithmetic sequence is an ordered list of numbers with a common difference (which is a constant) between successive terms. Consider the following list of numbers:

3 8 13 18 23 28 ...

This an example of an arithmetic sequence with a common difference of 5.

In this chapter we extend the work on sequences and series and look at sequences in which consecutive terms are connected by a constant multiplier which we call a **common ratio**.

Essential notes

Using notation covered in Core 1

$u_1 = 3$

$u_2 = 3 + 5 = 8$

$u_3 = 8 + 5 = 13$

Geometric sequences

Looking at the following two sequences we can see that in each case to get from one term to the next we multiply by a constant which is the common ratio:

1, 2, 4, 8, 16, ... has a common ratio 2

81, 27, 9, 3, 1, ... has a common ratio $\frac{1}{3}$

These are both examples of geometric sequences.

Essential notes

A geometric sequence is also called a **geometric progression,** often abbreviated to **G.P.**

Example

John is saving £5 000 in a savings plan which gives a constant annual tax free interest of 4%. How much will his savings be worth in 5 years time?

Answer

Step 1: Savings value end of year 1: $5000 \times 1.04 = 5200$

Step 2: Savings value end of year 2: $5200 \times 1.04 = 5408$

Step 3: Savings value end of year 3: $5408 \times 1.04 = 5624.32$

Step 4: Savings value end of year 4: $5624.32 \times 1.04 = 5849.29$

Step 5: Savings value end of year 5: $5849.29 \times 1.04 = 6083.26$

so John's savings have grown to £6083.26

Method notes

Interest of 4% means you earn $\frac{4}{100}$ on the amount you had in the ISA at the beginning of each year.

$5000 + \frac{4}{100} \times 5000$

$= 5000\left(1 + \frac{4}{100}\right) = 5000(1.04)$

which is the amount at the end of the first year.

Hence the 4% interest leads to the multiplier of 1.04. This type of interest is known as compound interest.

The pattern of savings at the end of each year £5200, £5408, £5624.32, £5849.29, £6083.26 is an example of a **geometric sequence** with first term 5000 and **common ratio** 1.04. To get from one term to the next we multiply by 1.04.

Definition

A **geometric sequence** has a common ratio between successive terms.
If the first term is a and the common ratio r the first five terms would be:
$a, ar, ar^2, ar^3, ar^4, \ldots$

The nth term is $u_n = ar^{n-1}$

Essential notes

$u_1 = ar^0$

$u_2 = ar^1$

$u_3 = ar^2$

The power of r is 1 less than the number of the term, that is $u_n = ar^{n-1}$.

The test for confirming a geometric sequence

- Consider any pair of consecutive terms: u_2 and u_3, u_3 and u_4 up to u_8 and u_9
- Divide each consecutive pair: $\frac{u_2}{u_1}, \frac{u_3}{u_2}, \frac{u_4}{u_3} \dots \frac{u_9}{u_8}$
- If the answer from the division is the same for consecutive each pair then the sequence is geometric and the constant answer found is the common ratio of the sequence.

Example

Which of the following are geometric sequences? For each geometric sequence found, write down the common ratio *r* and the *n*th term.

a) 5, 10, 20, 40, 80, ...

b) 100, 20, 4, 0.8, 0.16, ...

c) 2, 5, 7, 12, 19, 31, ...

d) 4, 2, 0, −2, −4, ...

e) 2, −2, 2, −2, 2, −2, ...

Answer

a) **Step 1**: Divide consecutive terms: $\frac{10}{5}, \frac{20}{10}, \frac{40}{20}$, all $= 2$ which is constant

Step 2: State that this is a geometric sequence with $r = 2$

Step 3: State the general term $u_n = 5 \times 2^{n-1}$

b) Using the method for (a) shows that this is a geometric sequence with $r = \frac{1}{5}$ and $u_n = 100 \times \left(\frac{1}{5}\right)^{n-1}$

c) Dividing out the consecutive terms shows that there is no common ratio so this is not a geometric sequence. Each term is the sum of the previous two terms. This is a special type of sequence and is called a Fibonacci sequence.

d) The consecutive terms do not have a common ratio but do have a common difference of −2 therefore this is an arithmetic sequence.

e) Using the method from (a) shows this is a geometric sequence with $r = -1$ and $u_n = 2 \times (-1)^{n-1}$

Example

Find the 10th term of the following geometric sequences:

a) 16, −8, 4, −2, 1, …

b) $\frac{x}{2}, \frac{x^2}{6}, \frac{x^3}{18}, \frac{x^4}{54}, \frac{x^5}{162}, \ldots$

Answer

a) **Step 1**: State $a = 16$ and $r = -\frac{1}{2}$

Method notes

$\frac{u_2}{u_1} = \frac{-8}{16} = -\frac{1}{2}$

Step 2: General term of a geometric sequence (or geometric progression, GP) is $u_n = ar^{n-1}$

Step 3: The 10th term is $n = 10$ so state the 10th term using $n = 10$ $a = 16$ and $r = -\frac{1}{2}$

so $u_{10} = 16 \times \left(-\frac{1}{2}\right)^9 = -0.03125$

b) **Step 1**: state $a = \frac{x}{2}$ and find r by dividing any consecutive pair,

e.g. $r = \frac{u_4}{u_3} = \frac{x^4}{54}$ divided by $\frac{x^3}{18} = \frac{x^4}{54} \times \frac{18}{x^3} = \frac{x}{3}$

Step 2: General term of a geometric sequence is $u_n = ar^{n-1}$ and the 10th term is $n = 10$ so state the 10th term using $n = 10$, $a = \frac{x}{2}$, and $r = \frac{x}{3}$

so $u_{10} = \frac{x}{2} \times \left(\frac{x}{3}\right)^9 = \frac{x^{10}}{39366}$

Example

In a geometric sequence, $u_3 = 3600$ and $u_5 = 1296$

a) Find the two possible values for the common ratio r and the value of the first term a.

b) Write down the first six terms for each of the two possible sequences.

c) How do the two sequences differ?

Answer

a) **Step 1**: $u_n = ar^{n-1}$ so

$u_3 = ar^2 = 3600$ (1)

$u_5 = ar^4 = 1296$ (2)

Step 2: (2) ÷ (1) $= \frac{ar^4}{ar^2} = r^2 = \frac{1296}{3600} = 0.36$ (so $r = \pm 0.6$)

Step 3: Put r^2 in (1) gives $a \times (0.36) = 3600$ so $a = \frac{3600}{0.36} = 10\,000$

b) The first six terms of the two sequences are:

$r = 0.6$: 10 000, 6000, 3600, 2160, 1296, 777.6

$r = -0.6$:10 000, −6000, 3600, −2160, 1296, −777.6,

c) The signs alternate in the second sequence.

Geometric series

In Core 1 we found an arithmetic series by adding the terms of an arithmetic sequence.

The series generated by adding the terms of a geometric sequence is called a **geometric series.**

Sum of a geometric series

The notation S_n is used to denote the sum of the first n terms of a geometric sequence.

$$S_n = a + ar + ar^2 + ar^3 + \dots\dots\dots ar^{n-2} + ar^{n-1}$$

To find a formula for the sum of a geometric series we need to carry out some algebraic manipulation.

$$S_n = a + ar + ar^2 + \dots + ar^{n-2} + ar^{n-1} \qquad (1)$$

Multiplying both sides of equation (1) by r, meaning multiply every term of the series, gives:

$$rS_n = ar + ar^2 + ar^3 + \dots + ar^{n-1} + ar^n \qquad (2)$$

We can simplify the working by taking equation (1) from equation (2) leading to: $rS_n - S_n = -a + ar^n = ar^n - a = a(r^n - 1)$

Factorising leads to:

$$S_n (r - 1) = a(r^n - 1)$$

so $\quad S_n = \dfrac{a(r^n - 1)}{r - 1}$

or we can rewrite this if we multiply numerator and denominator by -1 as:

$$S_n = \frac{a(1 - r^n)}{1 - r}$$

The general formula for the sum of the first n terms of a geometric series is

therefore $S_n = \dfrac{a(r^n - 1)}{r - 1}$ or $S_n = \dfrac{a(1 - r^n)}{1 - r}$

Exam tips

You must learn this proof for the sum of a geometric series.

You must also learn the two alternative versions of the general formula and when best to use them.

Example

For each of the following geometric series, write down the ith term u_i and the sum to n terms S_n.

In each case find the least value of n such that $S_n > 1000$

a) $a = 1, r = 3$

b) $250 + 200 + 160 + 128 + 102.4 + \dots$

c) $800 + 160 + 32 + 6.4 + 1.28 + \dots$

Method notes

Version 1: $S_n = \dfrac{a(r^n - 1)}{r - 1}$ to be used when $r > 1$

Version 2: $S_n = \dfrac{a(1 - r^n)}{1 - r}$ to be used when $r < 1$

Answer

a) **Step 1**: State the general term with $a = 1$ and $r = 3$ which is:

$$u_i = a \times r^{i-1} = 3^{i-1}$$

Step 2: as $r > 1$ use version 1 of the general formula so:

$$S_n = \frac{a(r^n - 1)}{r - 1} = \frac{3^n - 1}{3 - 1} = \frac{3^n - 1}{2}$$

Step 3: If $S_n > 1000$ then $\dfrac{3^n - 1}{2} > 1000$

$\Rightarrow 3^n - 1 > 2000 \Rightarrow 3^n > 2001$

Using trial and improvement:

$n = 5$ gives $3^5 = 243$;

$n = 6$ gives $3^6 = 729$;

$n = 7$ gives $3^7 = 2187$

So the least (or smallest) value of n such that $S_n > 1000$ is $n = 7$

Method notes

Trial and improvement, as its name implies, is when you **try** a particular number value to put into the formula. You then change or **improve** on this number value until you reach an answer which is > 2000

b) $250 + 200 + 160 + 128 + 102.4 + \ldots$ so $a = 250$, $r = 0.8$ (< 1)

Using the method from a):

$$u_i = a \times r^{i-1} = 250 \times 0.8^{i-1}$$

$$S_n = \frac{a(1 - r^n)}{1 - r}$$

$$= \frac{250(1 - 0.8^n)}{1 - 0.8} = \frac{1250(1 - 0.8^n)}{0.2} = 1250(1 - 0.8^n)$$

If $S_n > 1000$ then $1250(1 - 0.8^n) > 1000$

$\Rightarrow (1 - 0.8^n) > \dfrac{1000}{1250}$

$\Rightarrow (1 - 0.8^n) > 0.8$

$\Rightarrow 1 > 0.8 + 0.8^n$ so $1 - 0.8 > 0.8^n$ so $0.8^n < 0.2$

Using trial and improvement:

$n = 5$ gives $0.8^5 = 0.33$

$n = 7$ gives $0.8^7 = 0.21$

$n = 8$ gives $0.8^8 = 0.17$

Least value of n such that $S_n > 1000$ (which means $0.8^n < 0.2$) is $n = 8$

Method notes

Consider the function 0.2^n

If $n = 1$ it is 0.2

If $n = 2$ it is $(0.2)^2 = 0.04$

If $n = 3$ it is $(0.2)^3 = 0.008$

So as n increases 0.2^n gets smaller and we write $0.2^n \to 0$

As $S_n = 1000(1 - 0.2^n)$ this means that S_n approaches the value of 1000 written as $S_n \to 1000$ so S_n is never > 1000

c) $800 + 160 + 32 + 6.4 + 1.28 +$ so $a = 800$, $r = 0.2$ (< 1)

Using the method from a):

$$u_i = a \times r^{i-1} = 800 \times 0.2^{i-1}$$

$$S_n = \frac{a(1 - r^n)}{1 - r}$$

$$= \frac{800(1 - 0.2^n)}{1 - 0.2} = \frac{800(1 - 0.2^n)}{0.8} = 1000(1 - 0.2^n)$$

If $S_n > 1000$ then $1000(1 - 0.2^n) > 1000$ so $1 - 0.2^n > 1$

therefore $0.2^n < 0$ (so 0.2^n has to be negative).

Since 0.2^n is always positive for all values of n it means that there is no value of n for which $S_n > 1000$

Sum to infinity of a geometric series

The geometric series in above $800 + 160 + 32 + 6.4 + 1.28 + \ldots$ led to the interesting result that its sum never exceeded 1000. As the number of terms increased the series tended towards (or approached) the value 1000 but never reached it.

In such cases the number 1000 is called the **limit of the sum** or more often **the sum to infinity** of the series. We also say that this series is **convergent** because as the number of terms increases indefinitely the sum of the series approaches a particular value, in this case 1000

The shorthand way of writing this is $S_\infty = 1000$.

A series converges when the common ratio is in the range $-1 < r < 1$

This means that r^n gets smaller and smaller as n increases so using the sum for the first n terms of a geometric series that is $S_n = \dfrac{a(1 - r^n)}{1 - r}$ or $S_n = \dfrac{a(r^n - 1)}{r - 1}$, converges to a limit of $\dfrac{a}{1 - r}$.

More formally we write: if $-1 < r < 1$ then $r^n \to 0$ as $n \to \infty$ and the sum to infinity of a geometric series is $S_\infty = \dfrac{a}{1 - r}$

Essential notes

$-1 < r < 1$ is often written as $|r| < 1$ and we read this as the numerical value (or modulus) of r is less than 1

Not all geometric series are convergent. For example, given the series $1 + 3 + 9 + 27 + 81 + \ldots$

$a = 1$ $r = 3$ (>1) so $S_n = \dfrac{a(r^n - 1)}{r - 1}$ so $S_n = \dfrac{3^n - 1}{2}$

Using the formula with: $n = 1$ we get $S_1 = 1$

$n = 2$ we get $S_2 = 4$

$n = 3$ we get $S_3 = 13$

$n = 4$ we get $S_4 = 40$

$n = 10$ we get $S_{10} = 29\,524$

so the more terms we add together the larger the answer!

We can therefore conclude that the geometric series $1 + 3 + 9 + 27 + 81 + \ldots$ **diverges**.

More formally we write that when the common ratio $r > 1$ then r^n gets larger and larger as n increases, so the sum $S_n = \dfrac{a(r^n - 1)}{r - 1}$ increases indefinitely as n increases.

Stop and think 1

What happens to the sum to infinity of $S_n = \dfrac{a(r^n - 1)}{r - 1}$ *when* $r < -1$?

The test for deciding whether a geometric series converges or diverges is to find whether $|r| < 1$ or $|r| > 1$

Example

Determine which of the following geometric series converge to a limit, and in these cases find this limit (the sum to infinity).

a) $3 + 3^2 + 3^3 + 3^4 + \ldots$

b) $\left(\frac{1}{3}\right) + \left(\frac{1}{3}\right)^2 + \left(\frac{1}{3}\right)^3 + \left(\frac{1}{3}\right)^4 + \ldots$

c) $0.7^4 + 0.7^5 + 0.7^6 + 0.7^7 + \ldots$

d) $1 - 4 + 16 - 64 + 256 - \ldots$

Answer

a) $3 + 3^2 + 3^3 + 3^4 + \ldots$ has common ratio $r = 3$

Since $r > 1$ the series does not converge.

b) $\left(\frac{1}{3}\right) + \left(\frac{1}{3}\right)^2 + \left(\frac{1}{3}\right)^3 + \left(\frac{1}{3}\right)^4 + \ldots$ has common ratio $r = \left(\frac{1}{3}\right)$

Since $r < 1$ the series converges with a sum to infinity of

$$\frac{a}{1-r} = \frac{\frac{1}{3}}{1 - \frac{1}{3}} = \frac{\frac{1}{3}}{\frac{2}{3}} = \frac{1}{2}$$

c) $0.7^4 + 0.7^5 + 0.7^6 + 0.7^7 + \ldots$ has common ratio $r = 0.7$

Since $r < 1$ the series converges to the sum to infinity

$$\frac{a}{1-r} = \frac{0.7^4}{1 - 0.7} = \frac{0.7^4}{0.3} = 0.800\dot{3}$$

d) $1 - 4 + 16 - 64 + 256 - \ldots$ has common ratio $r = -4$.

Since $r < -1$ the series does not converge.

Problem solving with geometric sequences and series

Problems involving monetary growth, population change, bacterial growth and decay can often be solved using geometric sequences.

Example

A publisher keeps a record of the number of copies of a certain book that are sold each week. In the first week after publication 5000 copies were sold, and in the second week 4500 copies were sold. The publisher forecasts future sales by assuming that the copies sold each week will form a geometric sequence with first two terms 5000 and 4500.

Calculate the publisher's forecasts for

a) the number of copies sold in the 15th week after publication

b) the total number of copies sold during the first 15 weeks after publication

c) the maximum number of copies sold.

Answer

a) The assumed geometric sequence has first two terms 5000 and 4500

$\Rightarrow a = 5000$ and $r = \dfrac{4500}{5000} = 0.9\ (r < 1)$

$\Rightarrow u_n = 5000 \times 0.9^{n-1}$

Copies sold in the 15th week

$\Rightarrow u_{15} = 5000 \times 0.9^{14} = 1143$

which is the required forecast

b) the **total** number of copies sold during the first 15 weeks after publication is the **sum** of all the copies sold each week so:

$$S_{15} = 5000\frac{(1 - 0.9^{14})}{(1 - 0.9)} = 38\,561$$

c) Since $r = 0.9 < 1$ the series will converge and therefore the sum to infinity will be the maximum number of copies sold:

$$S_\infty = \frac{a}{1 - r} = \frac{5000}{1 - 0.9} = 50\,000$$

Method notes

The common ratio value is used throughout the question.

a) The copies sold in the 15th week will be given by u_{15}. The answer will need to be rounded to give a sensible result.

b) $r < 1$ so use version 2 of the formula.

The binomial expansion

You are familiar with the two formulae $(a + b)^2 = a^2 + 2ab + b^2$ and $(a - b)^2 = a^2 - 2ab + b^2$ when working with quadratic equations. In Core 2 we investigate expansions involving higher powers of n. The expansion of $(a + b)^n$ in terms of powers of a and powers of b is called the **binomial expansion**.

Example

Find expressions for $(a + b)^3$ and $(a + b)^4$ by expanding the brackets.

Answer

$(a + b)^3 = (a + b) \times (a + b) \times (a + b)$

$= (a^2 + 2ab + b^2) \times (a + b)$

$= (a^3 + 2a^2b + ab^2) + (a^2b + 2ab^2 + b^3)$

$= a^3 + 3a^2b + 3ab^2 + b^3 \qquad (1)$

$(a + b)^4 = (a + b) \times (a + b) \times (a + b) \times (a + b)$

so from (1) $= (a^3 + 3a^2b + 3ab^2 + b^3) \times (a + b)$

$= (a^4 + 3a^3b + 3a^2b^2 + ab^3) + (a^3b + 3a^2b^2 + 3ab^3 + b^4)$

$= a^4 + 4a^3b + 6a^2b^2 + 4ab^3 + b^4$

Method notes

$(a + b)^3 = (a + b)^2\,(a + b)$

Expand $(a + b)^2$ then multiply the result $(a^2 + 2ab + b^2)$ first by a then by b.

Finally collect together like terms.

If we continued the process in the example we would obtain

$(a + b)^5 = a^5 + 5a^4b + 10a^3b^2 + 10a^2b^3 + 5ab^4 + b^5$

$(a + b)^6 = a^6 + 6a^5b + 15a^4b^2 + 20a^3b^3 + 15a^2b^4 + 6ab^5 + b^6$

Essential notes

Index is another word for power.

Various patterns can be observed from these expansions as follows:

1. For each term of the expansion the sum of the indices of a and b is the same as the **index** of $(a+b)$.

   Consider $(a+b)^5 = a^5 + 5a^4b + 10a^3b^2 + 10a^2b^3 + 5ab^4 + b^5$

   The 2nd term in the expansion is $5a^4b$. The index of a^4 and the index of b^1 added together are 5 and the index of $(a+b)^5 = 5$

   The 4th term in the expansion is $10a^2b^3$. The index of a^2 and the index of b^3 added together is 5

   Similarly for the other terms in this expansion the sum of the indices of a and b is always 5

2. The first term in the expansion has a coefficient of 1 and the same index as $(a+b)$:

   $(a+b)^5 = a^5 + 5a^4b + \ldots$ (here the index = 5)

   $(a+b)^6 = a^6 + 6a^5b + \ldots$ (here the index = 6)

3. The coefficients of each term form a symmetric pattern, e.g.

   $(a+b)^5 = a^5 + 5a^4b + 10a^3b^2 + 10a^2b^3 + 5ab^4 + b^5$

   Coefficients: 1 5 10 10 5 1

   The coefficients of the increasing powers of $(a+b)$ form a pattern known as **Pascal's Triangle**:

Method notes

The mathematical pattern in Pascal's Triangle is that each line must start and end with 1

The rest of the numbers are found by adding the two numbers together in the line above.

$(a+b)^0$ 1

$(a+b)^1$ 1 1

$(a+b)^2$ 1 2 1

$(a+b)^3$ 1 3 3 1

1 4 6 4 1

1 5 10 10 5 1

1 6 15 20 15 6 1

Example

a) Use Pascal's triangle to expand $(a+b)^7$.

b) Find the first four terms of the expansion of $(x-2)^7$ in descending powers of x.

Answer

a) **Step 1**: The next line of Pascal's Triangle is

   1 7 21 35 35 21 7 1

   which gives the coefficients of the required expansion

   Step 2: Using the pattern described above when $n = 7$:

   $(a+b)^7 = a^7 + \mathbf{7}a^6b + \mathbf{21}a^5b^2 + \mathbf{35}a^4b^3 + \mathbf{35}a^3b^4 + \mathbf{21}a^2b^5 + \mathbf{7}ab^6 + b^7$

Essential notes

The terms written as a^7 and b^7 each have coefficient of 1

b) **Step 1**: Compare $(x-2)^7$ with $(a+b)^7$ so $a = x$ and $b = -2$

   Step 2: Using the pattern described above when $n = 7$

   $(x-2)^7 = x^7 + \mathbf{7}x^6 \times (-2) + \mathbf{21}x^5 \times (-2)^2 + \mathbf{35}x^4 \times (-2)^3 +$

   Step 3: Collect terms$(x-2)^7 = x^7 - 14x^6 + 84x^5 - 280x^4 + \ldots$

Example

The coefficient of x^2 in the expansion of $(2 + x)(4 + ax)^3$ is 144. Find the possible values of a.

Answer

Step 1: From Pascal's Triangle the coefficients of $(4 + ax)^3$ are 1, 3, 3, 1

Step 2: Using these coefficients and the pattern leads to:

$$(4 + ax)^3 = 4^3 + 3 \times 4^2(ax) + 3 \times 4(ax)^2 + (ax)^3$$

Step 3: Collecting like terms:

$$(4 + ax)^3 = 64 + 48ax + 12a^2x^2 + a^3x^3$$

Step 4: To find the value of a we now need to find all the terms in x^2 from $(2 + x)(4 + ax)^3$ so substitute the result in step 3 for $(4 + ax)^3$:

$$(2 + x)\ (64 + 48ax + 12a^2x^2 + a^3x^3)$$
$$= (2 \times 12a^2x^2) + (x \times 48ax)$$
$$= (24a^2 + 48a)x^2$$

Step 5: Using the information given about the coefficient of x^2 we find that

$$24a^2 + 48a = 144$$
$$\Rightarrow a^2 + 2a - 6 = 0$$
$$\Rightarrow \quad a = \frac{-2 \pm \sqrt{4 - 4(-6)}}{2}$$
$$\Rightarrow \quad a = \frac{-2 \pm \sqrt{4 + 24}}{2}$$
$$\Rightarrow \quad a = \frac{-2 \pm \sqrt{28}}{2}$$
$$\Rightarrow \quad a = \frac{-2 \pm 2\sqrt{7}}{2}$$
$$\Rightarrow \quad a = -1 \pm \sqrt{7}$$

Factorial notation

Pascal's Triangle provides a method of finding the coefficients in the expansion of $(a + b)^n$ for small positive integer values of n. For larger indices the method is too cumbersome.

A formula for the terms in Pascal's Triangle can be formed by considering the expansion of $(a + b)^n$ from first principles.

Consider again the expansion of $(a + b)^4$.

$$(a + b)^4 = (a + b) \times (a + b) \times (a + b) \times (a + b)$$
$$= a^4 + 4a^3b + 6a^2b^2 + 4ab^3 + b^4$$

The term in a^3b is formed by choosing a from any three of the brackets, and then multiplying these together with b from the remaining bracket.

This is similar to choosing 3 people from 4. Suppose that we label the people A, B, C and D the choices could be ABC, ABD, ACD, BCD so there are 4 ways of choosing 3 people where the order does not matter as ABC is the same choice as BAC

This number of ways **(4)** is the coefficient of a^3b in the expansion of $(a+b)^4$.

But if order matters then there are 24 ways of writing down the three letters from ABCD as shown below:

ABC	ACB	BAC	BCA	CAB	CBA
ABD	ADB	BAD	BDA	DAB	DBA
ACD	ADC	CAD	CDA	DCA	DAC
BCD	BDC	CBD	CDB	DBC	DCB

There are:

4 choices for the first letter: A, B, C, or D

3 choices for the second letter: e.g. if A is chosen as the first letter you can only choose from B, C, or D for the second letter

2 choices for the third letter e.g. if A and B are chosen as the first 2 letters you can only choose from C and D for the third letter.

Result 1: The total number of choices where the order matters is therefore $4 \times 3 \times 2$ which is the same as $4 \times 3 \times 2 \times 1$

Now for **each** of these three letter groupings chosen there are:

3 choices for the first letter (A, B or C)

2 for the second (because one letter has been chosen already)

1 for the third letter (because two letters have already been chosen from the original 3)

Result 2: The total number of ways of arranging any one of the 3 letter grouping is $3 \times 2 \times 1$ which is as follows:

ABC ACB BAC BCA CAB CBA

If we now use these two results we see that if *order does not matter*

a formula for choosing 3 people from 4 is $\dfrac{4 \times 3 \times 2 \times 1}{3 \times 2 \times 1}$

The expression above can be written using the factorial notation. The factorial notation is a shorthand way of writing multiplications such as:

$4 \times 3 \times 2 \times 1$ which is written as 4! and is pronounced '4 factorial'.

$6 \times 5 \times 4 \times 3 \times 2 \times 1$ which is written as 6! and is pronounced '6 factorial'

$3 \times 2 \times 1$ which is written as 3! and is pronounced '3 factorial'.

Looking at the pattern of these numbers we can see that each number multiplied is one less than the number before it and the last number of the multiplication must be 1.

We can also use letters to represent numbers in a factorial pattern so

$$n \text{ factorial} = n \times (n-1) \times (n-2) \times (n-3) \times \ldots \times 3 \times 2 \times 1$$

for positive integer values of n.

The formula for choosing 3 people from 4 can be written using the factorial notation:

$$\frac{4 \times 3 \times 2 \times 1}{3 \times 2 \times 1} = \frac{4!}{3!1!} = \frac{4!}{1! \times 3!}$$

These are written as ${}^4C_1$ where C stands for 'choose' so ${}^4C_2$ means the number of ways of choosing 2 objects from 4.

An alternative notation is $\binom{4}{1}$ and $\binom{4}{2}$ respectively.

Consider again the expansion of $(a+b)^4$:

$(a+b)^4 = (a+b) \times (a+b) \times (a+b) \times (a+b) = a^4 + 4a^3b + 6a^2b^2 + 4ab^3 + b^4$

The term in a^3b is formed by choosing a from any three of the brackets, choosing b from the one bracket not used to choose the three a terms and then multiplying the three chosen a terms together with b from the remaining bracket. So effectively you have chosen the b term from one of the four brackets.

From the earlier explanations and bringing this all together we get:

${}^4C_1$ or $\binom{4}{1}$ is the coefficient of a^3b in the expansion of $(a+b)^4$.

${}^4C_2$ or $\binom{4}{2}$ is the coefficient of a^2b^2 in the expansion of $(a+b)^4$.

More generally: the coefficient of a^rb^{n-r} in the expansion of $(a+b)^n$ is formed by choosing a from r brackets and b from $(n-r)$ brackets, which can be done in ${}^nC_r$ or $\binom{n}{r}$ ways:

$${}^nC_r = \binom{n}{r} = \frac{n!}{r!(n-r)!}$$

Definition of the binomial expansion as the series for $(a+b)^n$

Version 1

$(a+b)^n = {}^nC_0a^n + {}^nC_1a^{n-1}b + {}^nC_2a^{n-2}b^2 + {}^nC_3a^{n-3}b^3 + \ldots {}^nC_nb^n$

Version 2

$$(a+b)^n = \binom{n}{0}a^n + \binom{n}{1}a^{n-1}b + \binom{n}{2}a^{n-2}b^2 + \binom{n}{3}a^{n-3}b^3 + \ldots \binom{n}{n}b^n$$

If we apply this to the expansion of $(a+b)^4$ it may be easier to think of the number in the lower position of the two notations ${}^4C_2$ or $\binom{4}{2}$ being the power of the b term in the required term of the binomial expansion. So if we wanted the coefficient of a^2b^2 it would be ${}^4C_2$ or $\binom{4}{2}$.

If we wanted the coefficient of a^3b it would be ${}^4C_1$ or $\binom{4}{1}$.

Essential notes

It is important to recognise both factorial notations

Method notes

In the previous example when finding the coefficient of a^3b^1 in the expansion of $(a+b)^4 n = 4, r = 3$ so $n - r = 1$

Exam tips

You do not need to learn these formulae or how they are derived. They will be given to you for the examination. You do need to learn how to apply these formulae and to locate the correct button ${}^nC_r$ on your calculator to evaluate the coefficients or know how to calculate using factorials.

Example

Write down the expansion of the following:

a) $(x+2y)^5$

b) $(x-2)^6$

Method notes

In this example we show both notations. You only need to use one. Choose the one which you are most confident in using.

In either case make sure you use the correct button on your calculator.

Answer

a) **Step 1**: For $(x+2y)^5$ $n=5$ and using version 1 the coefficients are:

$${}^5C_0 = \frac{5!}{0!5!} = 1,\ {}^5C_1 = \frac{5!}{4!1!} = 5,\ {}^5C_2\frac{5!}{3!2!} = 10,\ {}^5C_3 = \frac{5!}{2!3!} = 10,$$

$${}^5C_4 = 5,\ {}^5C_5 = 1$$

Step 2: Apply the binomial expansion:

$$(x+2y)^5 = x^5 + 5x^4(2y) + 10x^3(2y)^2 + 10x^2(2y)^3 + 5x(2y)^4 + (2y)^5$$

$$= x^5 + 10x^4y + 40x^3y^2 + 80x^2y^3 + 80xy^4 + 32y^5$$

b) **Step 1**: For $(x-2)^6$ $n=6$ using version 2 the coefficients are:

$$\binom{6}{0} = \frac{6!}{6!0!} = 1,\ \binom{6}{1} = \frac{6!}{5!1!} = 6,\ \binom{6}{2} = \frac{6!}{4!2!} = 15,$$

$$\binom{6}{3} = \frac{6!}{3!3!} = 20,\ \binom{6}{4} = \frac{6!}{2!4!} = 15,\ \binom{6}{5} = \frac{6!}{1!5!} = 6,$$

$$\binom{6}{6} = \frac{6!}{0!6!} = 1$$

Step 2: apply the binomial expansion

$$(x-2)^6 = x^6 + 6x^5(-2) + 15x^4(-2)^2 + 20x^3(-2)^3 + 15x^2(-2)^4 + 6x(-2)^5 + (-2)^6$$

$$= x^6 - 12x^5 + 60x^4 - 160x^3 + 240x^2 - 192x + 64$$

Method notes

You only need to write down the x^5 term not the full expansion.

Example

Find the coefficient of x^5 in the expansion of $(2+3x)^{11}$.

Answer

The term involving x^5 is $C({}^{11}C_5)2^6(3x)^5 = \frac{11!}{6!5!}2^6 3^5 x^5$ so coefficient of x^5 is $462 \times 2^6 \times 3^5 = \mathbf{7\,185\,024}$

The binomial series

In your examination you are likely to be asked to expand a function to give a series in terms of powers of x. For expressions of the form $(1+x)^n$ and $(a+bx)^n$ the binomial expansion provides such a series.

$$(1+x)^n = \binom{n}{0}1^n + \binom{n}{1}1^{n-1}x + \binom{n}{2}1^{n-2}x^2 + \binom{n}{3}1^{n-3}x^3 + \binom{n}{4}1^{n-4}x^4 + \ldots$$

$$1 + nx + \frac{n(n-1)}{2!}x^2 + \frac{n(n-1)(n-2)}{3!}x^3 + \frac{n(n-1)(n-2)(n-3)}{4!}x^4 + \ldots + \ldots$$

This is an example of a **binomial series**.

Essential notes

Notice that in each term, the power of x is the same as the factorial denominator in the coefficient.

Example

Use the binomial series to find the first four terms of

a) $(1+x)^7$

b) $(1-2x)^6$

c) $(3+x)^5$

Answer

a) $(1+x)^7 = 1 + 7x + \frac{7\times 6}{2\times 1}x^2 + \frac{7\times 6\times 5}{3\times 2\times 1}x^3$

$= 1 + 7x + 21x^2 + 35x^3$

b) $(1-2x)^6 = 1 + 6(-2x) + \frac{6\times 5}{2\times 1}(-2x)^2 + \frac{6\times 5\times 4}{3\times 2\times 1}(-2x)^3$

$= 1 - 12x + 60x^2 - 160x^3$

c) $(3+x)^5 = 3^5 \times \left(1+\frac{x}{3}\right)^5$

$\left(1+\frac{x}{3}\right)^5 = 1 + 5\times\frac{x}{3} + \frac{5\times 4}{2}\times\left(\frac{x}{3}\right)^2 + \frac{5\times 4\times 3}{3\times 2\times 1}\times\left(\frac{x}{3}\right)^3$

$= 1 + \frac{5}{3}x + \frac{10}{9}x^2 + \frac{10}{27}x^3$

$(3+x)^5 = 3^5\times\left(1+\frac{x}{3}\right)^5 = 243 + 405x + 270x^2 + 90x^3$

Method notes

a) This is straight forward as you can just apply the expansion with $n = 7$ Only list the first four terms.

b) Care is needed here. $(1-2x)^6 = (1+(-2x))^6$

Compare with the general binomial series expansion: 'x' is replaced by '$-2x$' and $n = 6$

c) Care is needed here. $(3+x)^5 = \left(3\left(1+\frac{x}{3}\right)\right)^5 =$

$3^5 \times \left(1+\frac{x}{3}\right)^5$

Comparing with the general binomial series expansion 'x' is replaced by '$\frac{x}{3}$' and $n = 5$

Example

a) Use the binomial series to find the first four terms of $(1+x)^{10}$.

b) Hence find an approximation for 1.04^{10} correct to three significant figures.

c) Hence find an approximation for 0.98^{10} correct to three significant figures.

Answer

a) This is straight forward with $n = 10$. Apply the expansion formula:

$$(1+x)^{10} = 1 + 10x + \frac{10 \times 9}{2 \times 1}x^2 + \frac{10 \times 9 \times 8}{3 \times 2 \times 1}x^3 + \ldots$$

$$(1+x)^{10} = 1 + 10x + 45x^2 + 120x^3 + \ldots$$

b) *Hence* means we should use the answer from a) so:

Step 1: Rewrite 1.04 as $1 + 0.04$

Step 2: Compare this with the expansion in a) and replace 'x' by '0.04':

$$1.04^{10} = (1 + 0.04)^{10}$$

Step 3: Use the expansion formula to give:

$$\begin{aligned} 1.04^{10} &= 1 + 10 \times 0.04 + 45(0.04)^2 + 120(0.04)^3 + \ldots \\ &= 1 + 0.4 + 0.072 + 0.00768 + \ldots \\ &= 1.47968 + \ldots \\ &= 1.48 \text{ correct to three significant figures} \end{aligned}$$

Only the first three terms are required to give an accuracy to three significant figures.

c) *Hence* means we should use the answer from a) so:

Step 1: Rewrite 0.98 as $1 - 0.02$

Step 2: Compare this with the expansion in a) and replace 'x' by '-0.02'

$$0.98^{10} = (1 + (-0.02)^{10}$$

Step 3: Use the expansion formula to give $0.98^{10} = 1 + 10 - (-0.02) + 45(-0.02)^2 + 120(-0.02)^3 + \ldots = 1 - 0.2 + 0.018 - 0.00096 + \ldots = 0.81704 + \ldots = 0.817$ correct to three significant figures

In this case the first four terms are required to give an accuracy to three significant figures.

Stop and think answers

1 Using $S_n = \frac{a(r^n - 1)}{r - 1}$ *take a value of* $r < -1$

$r = -2$ *and increasing values of* n

$n = 1 \quad r = -2 \qquad S_1 = \frac{a(-2 - 1)}{-2 - 1} = a$

$n = 2 \quad r^2 = (-2)^2 = 4 \qquad S_2 = \frac{a(4 - 1)}{-3} = -a$

$n = 3 \quad r^3 = (-2)^3 = -8 \qquad S_3 = \frac{a(-8 - 1)}{-3} = 3a$

$n = 4 \qquad S_4 = -5a$ *and so on*

This shows an oscillating sequence of answers (they alternate between positive and negative) and the magnitudes of the answers are increasing so the series would not converge. This means that you cannot give an answer for the sum to infinity.

Solution of triangles

Solving problems involving angles and side lengths in right angled triangles was covered in the GCSE course.

Example

Calculate the unknown angle and side lengths in the triangles shown in Figure 4.1. In each case calculate the area of the triangle.

Fig. 4.1

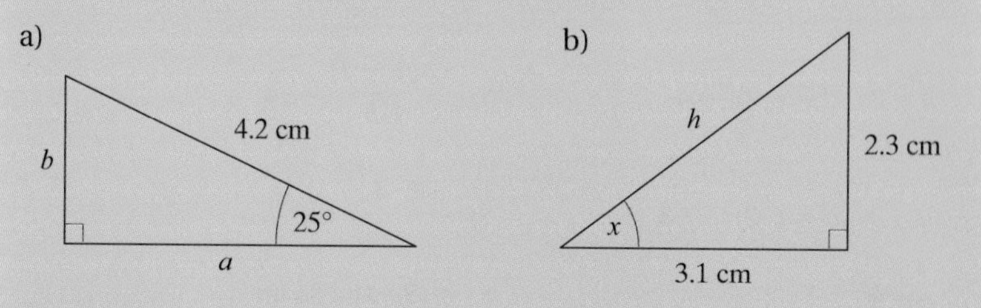

Essential notes

In any right angled triangle the side opposite the right angle is called the hypotenuse. The three trigonometric ratios are:

$$\text{Sine} = \frac{\text{opposite}}{\text{hypotenuse}}$$

$$\text{Cosine} = \frac{\text{adjacent}}{\text{hypotenuse}}$$

$$\text{Tangent} = \frac{\text{opposite}}{\text{adjacent}}$$

Area of a triangle = $\frac{1}{2} \times \text{base} \times \text{height}$ where the base and height are at 90° to each other. You should know this formula.

Answer

a) $\sin 25° = \frac{b}{4.2}$

$b = 4.2 \times \sin 25° = 1.775 \text{ cm}$

$\cos 25° = \frac{a}{4.2}$

$a = 4.2 \times \cos 25° = 3.806 \text{ cm}$

$\text{area} = \frac{1}{2} \times a \times b = \frac{1}{2} \times 3.806 \times 1.775 = 3.38 \text{ cm}^2$

b) Pythagoras' Theorem gives

$h^2 = 3.1^2 + 2.3^2 = 14.9$

$h = 3.86 \text{ cm}$

$\tan x = \frac{2.3}{3.1} = 0.7419$

$x = 36.6°$

$\text{area} = \frac{1}{2} \times 3.1 \times 2.3 = 3.57 \text{ cm}^2$

When the triangle is NOT right angled, we need the **sine** and **cosine rules** to calculate the length of sides and the size of angles. A standard convention of labelling the sides is adopted for such problems as shown in Figure 4.2.

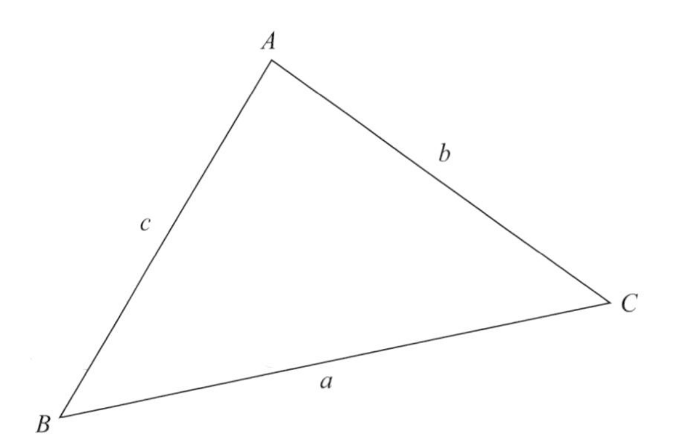

Fig. 4.2
The standard convention for labelling triangles.

Lower case letters a, b and c are used for the side lengths.

a is opposite $\angle A$, b is opposite $\angle B$, c is opposite $\angle C$.

Finding the sine rule formulae

Draw the line AM from A perpendicular to the side BC, as shown in Figure 4.3 and let h be the length of AM.

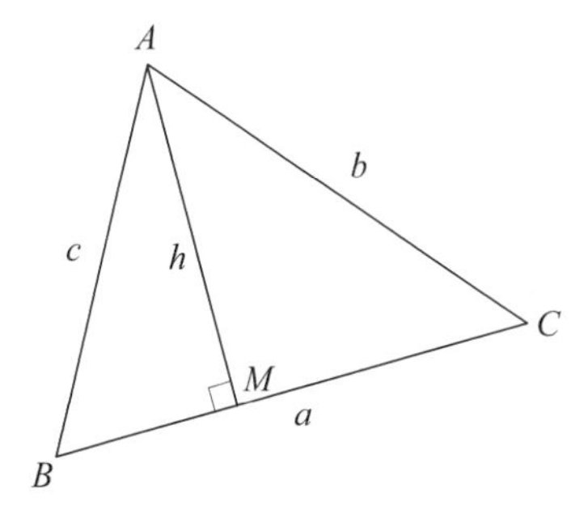

Fig. 4.3

Step 1: In $\Delta\ ABM$: $\frac{h}{c} = \sin B$ so $h = c \sin B$

Step 2: In $\Delta\ AMC$: $\frac{h}{b} = \sin C$ so $h = b \sin C$

Step 3: Equating the result for h in step 1 and step 2 gives

$c \sin B = b \sin C$

so $\frac{c}{\sin C} = \frac{b}{\sin B}$

Draw the perpendicular from B to AC as shown in Figure 4.4.

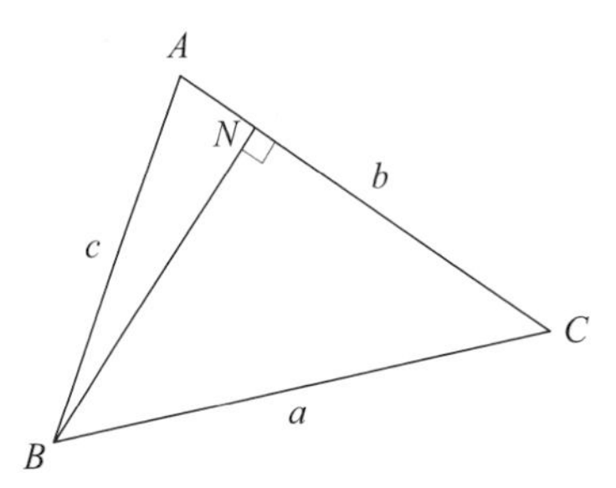

Fig. 4.4

Step 4: In $\Delta\ ABN$: $\frac{BN}{c} = \sin A$ so $BN = c \sin A$

Step 5: In $\Delta\ CBN$: $\frac{BN}{a} = \sin C$ so $BN = a \sin C$

Step 6: Equating the result for BN in step 4 and step 5 gives

$c \sin A = a \sin C$ so

$\frac{c}{\sin C} = \frac{a}{\sin A}$

Equating $\frac{c}{\sin C}$ from step 3 and step 6 gives

$\frac{a}{\sin A} = \frac{b}{\sin B} = \frac{c}{\sin C}$ or $\frac{\sin A}{a} = \frac{\sin B}{b} = \frac{\sin C}{c}$

These are the two versions of the sine rule.

Exam tips

You must learn the sine rule formulae. They will **not** be on the formula sheet provided in the examination. You must also learn which version to use when solving triangles.

Example

In Figure 4.5 below find the values of a and b.

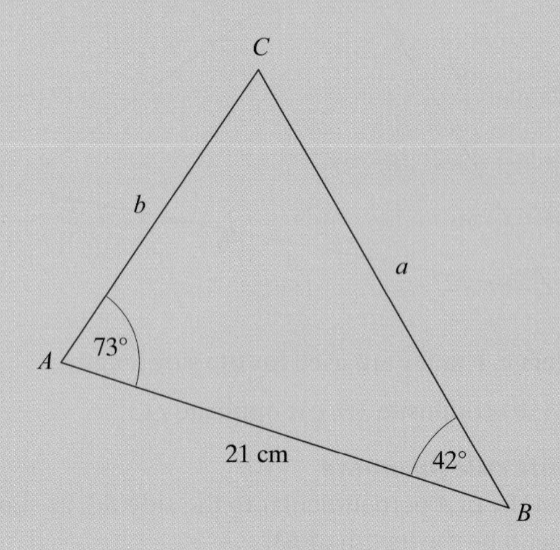

Fig. 4.5

Answer

The angle at $C = 180° - 73° - 42° = 65°$

The sine rule gives $\dfrac{a}{\sin 73°} = \dfrac{b}{\sin 42°} = \dfrac{21}{\sin 65°}$

$$\Rightarrow \frac{a}{\sin 73°} = \frac{21}{\sin 65°}$$

$$\Rightarrow \quad a = \frac{21 \times \sin 73°}{\sin 65°}$$

$$= 22.2 \text{ cm (3 s.f.)}$$

Similarly: $\dfrac{b}{\sin 42°} = \dfrac{21}{\sin 65°}$

$$\Rightarrow b = \frac{21 \times \sin 42°}{\sin 65°}$$

$$= 15.5 \text{ cm (3 s.f.)}$$

Method notes

List the information given using conventional labelling so $c = 21$

$\angle A = 73°$ $\angle B = 42°$.

Use angles in a triangle add to 180° to find $\angle C$.

To find side lengths use version 1 of the sine rule:

$$\frac{a}{\sin A} = \frac{b}{\sin B} = \frac{c}{\sin C}$$

Example
In Figure 4.6 below find the size of angle C.

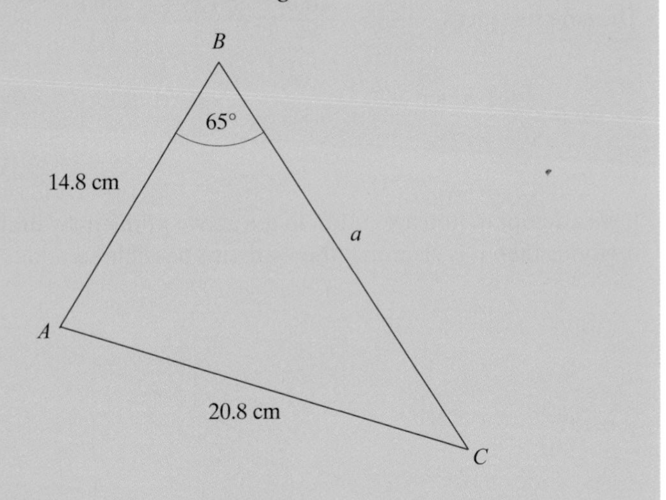

Fig. 4.6

Answer

The sine rule gives $\frac{\sin C}{c} = \frac{\sin B}{b}$ so $\frac{\sin C}{14.8} = \frac{\sin 65°}{20.8}$

$$\Rightarrow \sin C = \frac{14.8 \times \sin 65°}{20.8} = 0.6449$$

$$\Rightarrow \quad \angle C = 40.2°$$

Method notes

List the information given using conventional labelling so $b = 20.8$ $c = 14.8$ $\angle B = 65°$.

To find the size of angles use version 2 of the sine rule: $\frac{\sin A}{a} = \frac{\sin B}{b} = \frac{\sin C}{c}$.

In some problems there are two solutions for the missing angle. This is likely to occur when the angle you have found from the sine rule is larger than the given angle so that you can draw two possible triangles from the given data.

Example
In the triangle illustrated in Figure 4.7 below find the possible sizes of angle B.

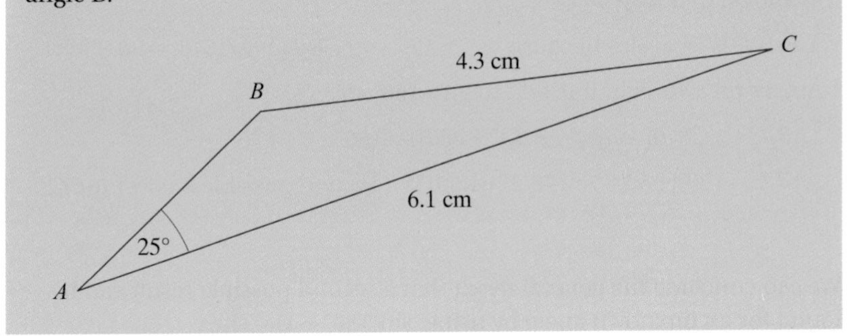

Fig. 4.7

Continued on the next page

Method notes

List the information given using conventional labelling so $a = 4.3$ $b = 6.1$ $\angle A = 25°$

To find angles use

$$\frac{\sin A}{a} = \frac{\sin B}{b} = \frac{\sin C}{c}.$$

Answer

The sine rule gives $\frac{\sin A}{a} = \frac{\sin B}{b}$ so $\frac{\sin 25°}{4.3} = \frac{\sin B}{6.1}$

so $\sin B = \frac{6.1 \times \sin 25°}{4.3} = 0.5995$

so $\angle B = 36.8°$

If we attempt to find a solution to the above problem by drawing methods then it is clear that there are two possible solutions for $\angle B$.

Fig.4.8

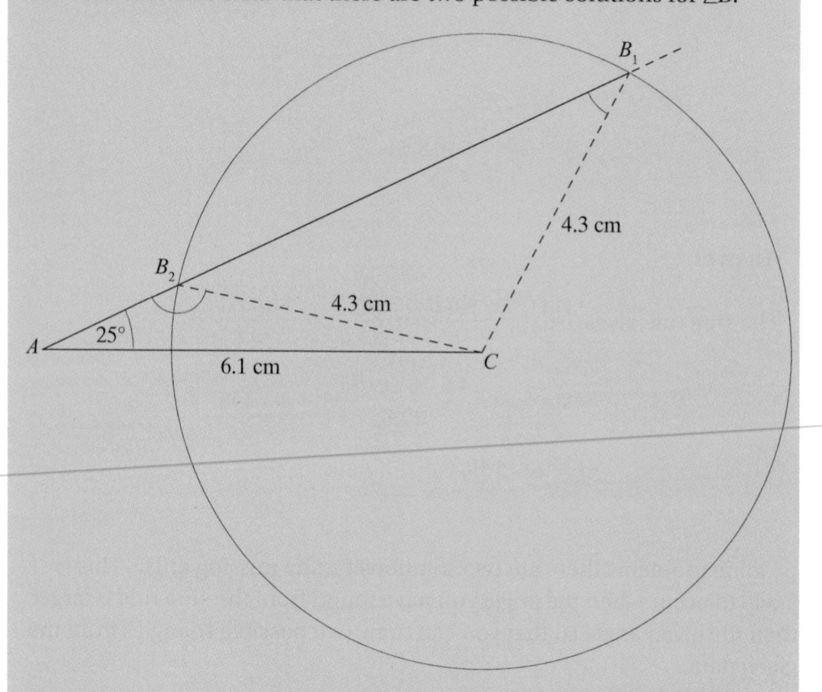

Method notes

First draw a line of length 6.1 cm.

Then draw a line through A at 25° to AC.

Now with a compass point at C draw a circle of radius 4.3 cm.

It cuts the line AB in two places at B_1 and B_2 giving two solutions to the problem for $\angle B$.

Essential notes

An isosceles triangle has two sides of equal length. Base angles of an isosceles triangle are also equal.

Method notes

When finding angles using the sine rule your calculator will give an acute angle. Remember there is a second (obtuse) angle which is equal to 180° – the acute angle.

The relationship between the two possible solutions for $\angle B$ becomes clear if we look at the two triangles AB_2C and AB_1C.

We have found that one solution is $\angle B = 36.8°$ which is an acute angle as shown in $\Delta\, AB_1C$.

$\Delta\, B_2B_1C$ is isosceles because $B_2C = CB_1$ so $\angle CB_1B_2 = \angle CB_2B_1 = 36.8°$.

Angles on a straight line add to give 180° so:

$\angle AB_2C + \angle CB_2B_1 = 180°$ $\angle AB_2C + 36.8° = 180°$

$\angle AB_2C = 180° - 36.8° = 143.2°$ which is a second possible answer for $\angle B$.

We can conclude the general result that a second possible result can be found for an unknown angle by using: $\sin(180° - x) = \sin x$

This work will be developed later in the chapter.

Finding the cosine rule formulae

In Figure 4.9 below, the line BM is perpendicular to the side AC and has length h. The three sides and the three angles of the triangle are labelled in the conventional way.

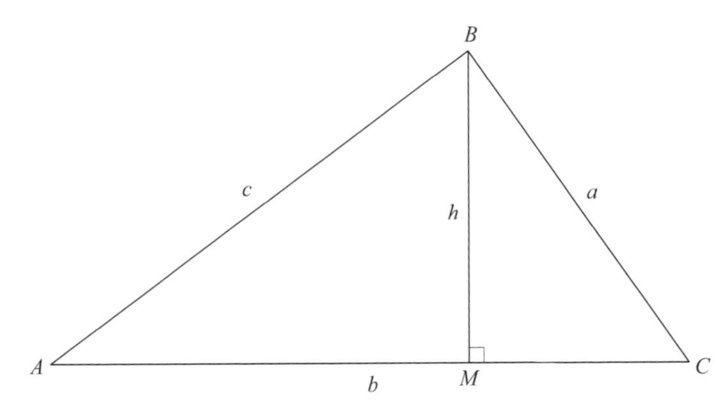

Fig. 4.9

To find the cosine rule:

Step 1: In $\Delta\, AMB$: $\frac{AM}{c}$ cos A so $AM = c \cos A$

Step 2: $AC = AM + MC$ and $AC = b$ so from step 1 $b = c \cos A + MC$ therefore $MC = b - c \cos A$

Step 3: Apply Pythagoras' theorem to $\Delta\, ABM$: $h^2 = c^2 - AM^2$ so

$$h^2 = c^2 - (c \cos A)^2 = c^2 - c^2 \cos^2 A$$

Step 4: Apply Pythagoras' theorem to $\Delta\, BCM$:

$$h^2 = a^2 - CM^2 = a^2 - (b - c \cos A)^2$$

$$\Rightarrow h^2 = a^2 - (b^2 - 2bc \cos A + c^2 \cos^2 A)$$

$$= a^2 - b^2 + 2bc \cos A - c^2 \cos^2 A$$

Step 5: Equating the two expressions for h^2 from step 3 and step 4 gives

$$c^2 - c^2 \cos^2 A = a^2 - b^2 + 2bc \cos A - c^2 \cos^2 A$$

$$\Rightarrow \quad c^2 = a^2 - b^2 + 2bc \cos A$$

This gives:

$$a^2 = b^2 + c^2 - 2bc \cos A$$

which is one version of the cosine rule.

The formula for a^2 involves the side lengths b and c and the cosine of angle A which is the angle between the sides b and c.

By similar approaches we could develop the other two versions of the cosine rule:

$$b^2 = a^2 + c^2 - 2ac \cos B$$

(using a, c and $\angle B$ which is the angle between the sides a and c)

$$c^2 = a^2 + b^2 - 2ab \cos C$$

(using a, b and $\angle C$ which is the angle between the sides a and b)

Method notes

Using conventional labelling: $BC = a \quad AC = b \quad AB = c$.

Method notes

Notice that if $\angle A = 90°$ then $\cos A = 0$ and the cosine rule shows $a^2 = b^2 + c^2$ which is Pythagoras' Theorem for a right angled triangle.

Exam tips

You do not need to learn the cosine rule but you must learn when and how to apply it for solving triangles.

From each version of the cosine rule we can see that if you are given two sides and the angle between those two sides then you can find the length of the unknown side.

Example

In the triangle shown in Figure 4.10, calculate the length of the third side AC.

Fig. 4.10

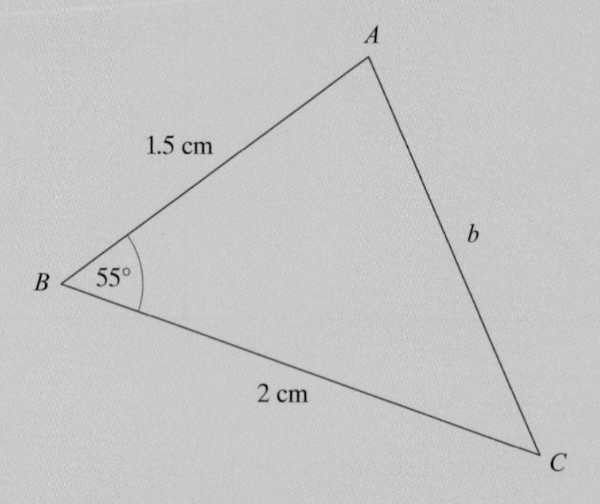

Method notes

List the information given using conventional labelling:

$a = 2$

$c = 1.5$

$\angle B = 55°$

This shows that we know the length of two sides of the triangle and the angle between those sides so we use the appropriate version of the cosine rule to find b.

Answer

Step 1: Identify from the information given which version of the cosine rule to use which is

$b^2 = a^2 + c^2 - 2ac \cos B$

Step 2: Substitute the values for a, c and $\angle B$ to give

$b^2 = 2^2 + 1.5^2 - 2 \times 2 \times 1.5 \cos 55°$ so $b^2 = 2.81$

Step 3: Evaluate b to give $b = 1.68$ cm which is the length of the side AC.

Example

In the triangle shown in Figure 4.11, calculate the size of the smallest angle.

Fig. 4.11

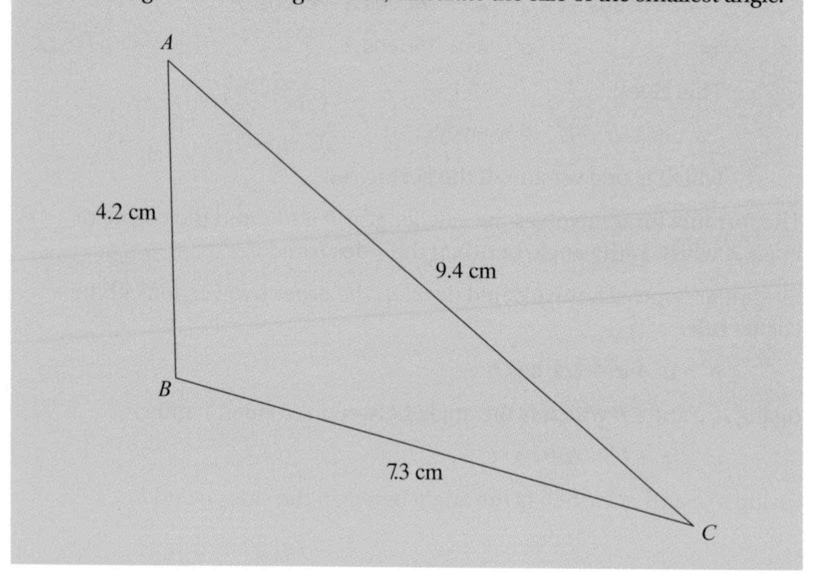

Answer
The smallest angle is opposite the smallest side c so we need to find $\angle C$.

Step 1: Use the rearranged cosine rule

$$\cos C = \frac{a^2 + b^2 - c^2}{2ab}$$

Step 2: Substitute the values for *a*, *b* and *c* in step 1 gives

$$\cos C = \frac{7.3^2 + 9.4^2 - 4.2^2}{2 \times 7.3 \times 9.4} = 0.9036$$

so $\angle C = 25.4°$ which is the smallest angle.

Method notes

List the information given using conventional labelling:

$a = 7.3$

$b = 9.4$

$c = 4.2$

The 3 versions of the cosine rule can be rearranged algebraically to find cos *A*, cos *B* or cos *C*. In this question we need to find cos *C* so use:

$c^2 = a^2 + b^2 - 2ab\cos C$

so $c^2 + 2ab\cos C = a^2 + b^2$

so $2ab\cos C = a^2 + b^2 - c^2$

Dividing both sides by 2 *ab*:

$$\cos C = \frac{a^2 + b^2 - c^2}{2ab}$$

In some problems you may need to use both the sine rule and the cosine rule. The following example illustrates this.

Example
Find angles *x* and *y* and sides *BC* and *AD* in Figure 4.12 below.

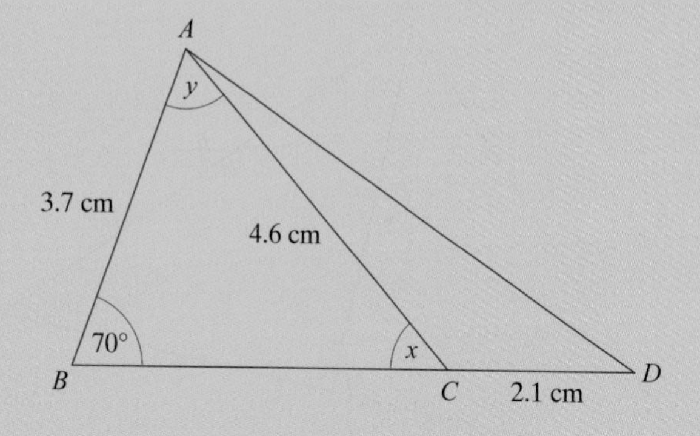

Fig. 4.12

Answer
Step 1: From ΔABC find angle *x* using the sine rule

$$\frac{\sin x}{3.7} = \frac{\sin 70°}{4.6}$$

$$\sin x = \frac{3.7 \times \sin 70°}{4.6} = 0.7558$$

$$x = 49.1°$$

Step 2: Angles in a triangle add to give 180° so in ΔABC:

$y = 180° - 70° - 49.1°$ so $y = 60.9°$

Method notes

The two triangles, ΔABC and ΔABD must be used to answer this question.

In ΔABC: $b = 4.6$ $c = 3.7$, $\angle B = 70°$, $\angle C = x°$ and $\angle A = y°$.

In ΔABD: $d = 3.7$, $\angle B = 70°$.

Step 3: Use the sine rule in ΔABC to find the length BC:

$$\frac{BC}{\sin y} = \frac{4.6}{\sin 70°}$$

Step 4: Substitute in step 3 for $y = 60.9°$ so $BC = \frac{4.6 \times \sin 60.9}{\sin 70°} = 4.28$

Step 5: $BD = BC + CD$

From step 4 $BC = 4.28$ therefore $BD = 4.28 + 2.1 = 6.38$

Step 6: Use the cosine rule in ΔABD to find AD so

$$AD^2 = AB^2 + BD^2 - 2 \times AB \times BD \cos 70°$$

$$= 3.7^2 + 6.38^2 - 2 \times 3.7 \times 6.38 \cos 70°$$

$$= 38.25$$

$$AD = 6.18$$

Method notes

In ΔABD we now know that $d = 3.7$, $a = 6.38$ and $\angle B = 70°$. This means we know the lengths of two sides and the angle between them so use the cosine rule.

Area of a triangle

Fig. 4.13

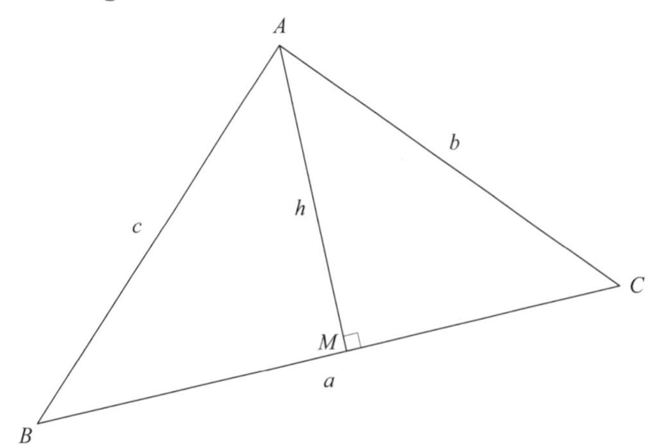

Essential notes

The sides of the triangle are labelled in the conventional way as explained at the beginning of this chapter.

$BC = a$, $AC = b$, $AB = c$.

The area of triangle ABC is $\frac{1}{2} \times a \times h$ (1)

where h is the length of the perpendicular from A to BC as shown in Figure 4.13.

Using the trigonometry of right angles triangles in ΔABM

$$\sin B = \frac{h}{c} \text{ so } h = c \times \sin B \qquad (2)$$

Substituting the value of h from equation (2) into statement (1) gives:

$$\text{area of } \Delta ABC = \frac{1}{2} a\, c \sin B$$

(using a, c and $\angle B$ which is the angle between a and c)

A similar approach using ΔACM gives:

$$\text{area of } \Delta ABC = \frac{1}{2} a\, b \sin C$$

(using a, b and $\angle C$ which is the angle between a and b)

Drawing the perpendicular from B to AC gives:

$$\text{area of } \Delta ABC = \frac{1}{2} b\, c \sin A$$

(using b, c and $\angle A$ which is the angle between b and c)

These formulae are used when you know the lengths of two sides in the triangle and the angle between these sides.

Exam tips

You must learn the 3 versions of the area formula: they will not be given on the formula sheet provided in the examination. You must also learn how to apply these formulae.

Example

In triangle ABC, $AB = 4.7$ cm, $AC = 7.3$ cm and angle $BAC = 70°$. Calculate the area of the triangle.

Answer

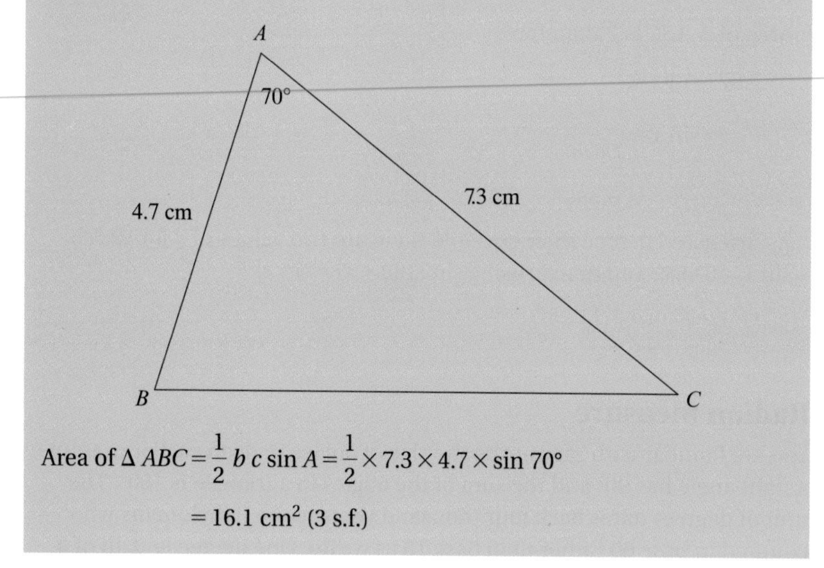

Fig. 4.14

$$\text{Area of } \Delta ABC = \frac{1}{2} b\, c \sin A = \frac{1}{2} \times 7.3 \times 4.7 \times \sin 70°$$

$$= 16.1 \text{ cm}^2 \text{ (3 s.f.)}$$

Method notes

Draw the triangle as shown in Figure 4.14 using the information given. List the information:

$b = 7.3$

$c = 4.7$

$\angle A = 70°$

Use the appropriate version of the area formula.

Stop and think 1

By drawing a perpendicular from B to AC in a general triangle ABC (not a right angled triangle) explain how to obtain the result:

area of a triangle ABC $= \frac{1}{2}$ *bc sin A.*

Example

The triangle shown in Figure 4.15 below has an area of 7 cm^2. Calculate the possible size(s) of angle x.

Fig. 4.15

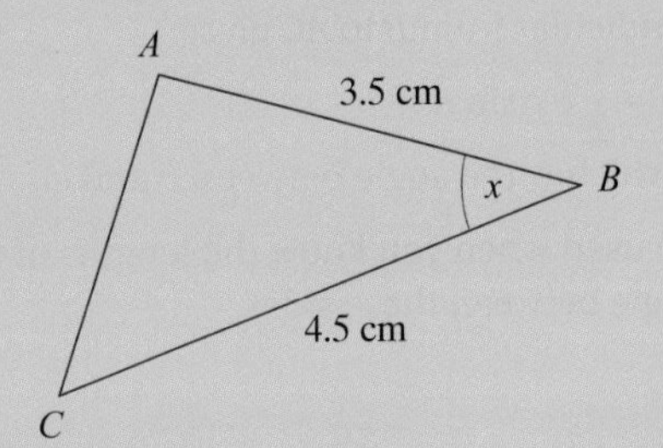

Method notes

List the information given:

$a = 4.5$

$c = 3.5$

area of triangle = 7

$\angle B = x°$

Use the appropriate version of the area formula.

In general $\sin(180° - x) = \sin x$.

Your calculator will give an acute angle between 0° and 90°.

A second (obtuse) angle is 180° minus this acute angle.

Answer

Area of $\Delta\ ABC = \frac{1}{2} ac \sin B = \frac{1}{2} \times 4.5 \times 3.5 \times \sin x$

$= 7.875 \sin x$

Area of $\Delta\ ABC$ is 7 therefore:

$7.875 \sin x = 7$

$\Rightarrow \sin x = \frac{7}{7.875}$

$= 0.8889$

As discussed in an earlier example there are two values of x for which $\sin x = 0.8889$ and hence, using $\sin (180 - x) = \sin x$:

$x = 62.7°$ or $x = 117.3°$

Essential notes

There are 24 factors of 60. There are 9 factors of 100.

There is a decimal system which divides a right angle into 100 divisions. The international standard symbol for this unit is the '**gon**'.

100 gons = 90°.

You may find this unit on a calculator and written as a 'grad'.

Radian measure

You are familiar with measuring the size of angles in degrees. For example, a right angle has 90° and the sum of the angles in a triangle is 180°. The unit of degrees dates back four thousand years to the Babylonians who counted in base 60 rather than base 10 as we do. One degree is $\frac{1}{360}$th of a complete revolution.

One advantage of the Babylonian system over the decimal system is that 60 has many factors so that we can divide a circle into many parts which have an integer number of degrees e.g. 4 angles of 90°; 6 angles of 60°; 9 angles of 40° and so on. Try this with a decimal system based on 100 divisions of a circle. A right angle would have 25 units – not so bad! Angles in an equilateral triangle would have $16\frac{2}{3}$ units – not so good!

An alternative system of units for measuring angles, called **radians** is based on the length of an arc of a circle.

Imagine you are riding on a big wheel of radius 8 m

(Figure 4.16). There are 12 equally spaced pods altogether.

In one revolution a pod will travel the circumference ($2\pi r$)

of the wheel, a distance of $2\pi \times 8 = 16\pi \approx 50.27$ m.

At the start of a ride the wheel stops at each pod so that people can get in. Between stops each pod travels $\frac{1}{12}$ of the circumference, a distance of $\frac{1}{12} \times 16\pi = \frac{4}{3}\pi \approx 4.19$ m and the wheel turns through $30°$ $\left(\frac{1}{12} \times 360°\right)$.
Therefore when the wheel travels through 30° it travels a length of 4.19 m along the circumference (known as the arc length). If a pod travels a distance of 8 metres (the same distance as the radius) the angle it turns through is $\frac{8}{4.19} \times 30° = 57.3°$.

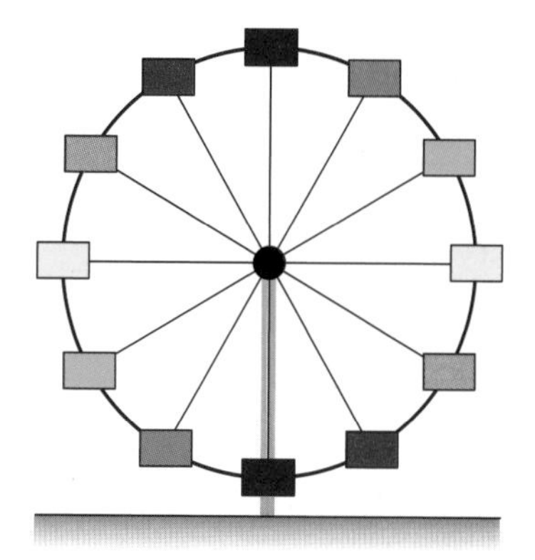

Fig. 4.16

This angle (57.3°) is called one **radian** because the arc length is the same distance as the radius.

Definition

One **radian** (see Figure 4.17) is defined to be the angle subtended (formed) at the centre of a circle of radius r by an arc of length r.

The symbol for radian measure is ^{c} so 2^c means 2 radians.

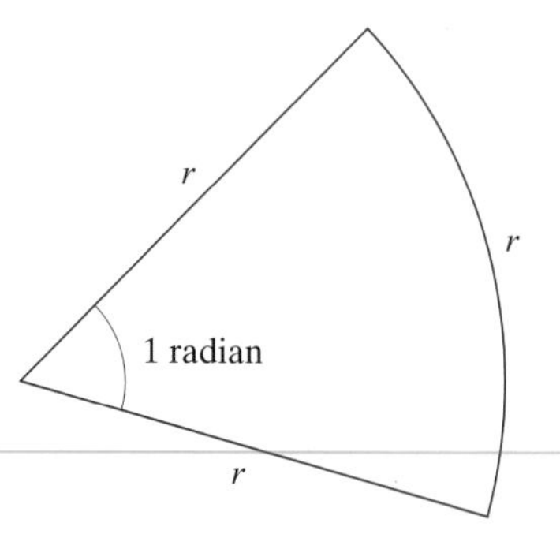

Fig. 4.17

Converting between degrees and radians

When the wheel turns through $x°$ the pod has travelled $\frac{x}{360}$ of a revolution which is arc length $\frac{x}{360} \times 16\pi$ (metres).

When this arc length is 8 metres the angle subtended is defined as 1 radian so $\frac{x}{360} \times 16\pi = 8$ so $x = \frac{180}{\pi} = 1$ radian.

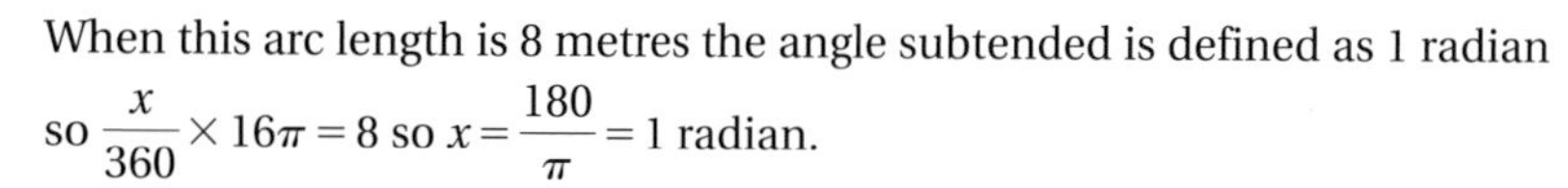

This provides a conversion relationship between degrees and radians.

$$1 \text{ radian} \equiv \frac{180}{\pi} \text{ degrees}$$

$$1 \text{ degree} \equiv \frac{\pi}{180} \text{ radians}$$

Example

Convert these angles from degrees to radians.

a) 30° b) 79° c) 300°

Answer

a) $1° \equiv \frac{\pi}{180}$ radians so $30° \equiv \frac{30}{180}\pi \equiv \frac{\pi}{6}$ radians $\equiv 0.524^c$

b) $1° \equiv \frac{\pi}{180}$ radians so $79° \equiv \frac{79}{180}\pi$ radians $\equiv 1.38^c$

c) $1° \equiv \frac{\pi}{180}$ radians so $300° \equiv \frac{300}{180}\pi = \frac{5}{3}\pi$ radians $\equiv 2.62^c$

Essential notes

These angles are all given in radians so you should change your calculator to 'RAD' mode. Use the sin, cos and tan keys in the usual way.

Essential notes

π radians $\equiv 180°$

An angle measuring 1 radian is 57.3° (just under 60°).

Method notes

To convert from degrees to radians multiply by $\frac{\pi}{180}$ using the π key on your calculator to work out the answers.

Method notes

To convert radians to degrees multiply by $\frac{180}{\pi}$ using the π key on your calculator to work out the answers.

Example

Convert these angles from radians to degrees.

a) $\frac{\pi^c}{5}$ b) $\frac{3\pi^c}{2}$ c) 0.52^c

Answer

a) $1^c \equiv \frac{180}{\pi}$ degrees so $\frac{\pi^c}{5} \equiv \frac{\pi}{5} \times \frac{180}{\pi} \equiv 36°$

b) $1^c \equiv \frac{180}{\pi}$ degrees so $\frac{3\pi^c}{2} \equiv \frac{3\pi}{2} \times \frac{180}{\pi} \equiv 270°$

c) $1^c \equiv \frac{180}{\pi}$ degrees so $0.52^c \equiv 0.52 \times \frac{180}{\pi} \equiv 29.8°$

Essential notes

Remember to check the mode of your calculator and use degrees or radians as required in the question.

You should become familiar with the following common angles in radians.

30°	45°	60°	90°	120°	135°	150°	180°
$\frac{\pi^c}{6}$	$\frac{\pi^c}{4}$	$\frac{\pi^c}{3}$	$\frac{\pi^c}{2}$	$\frac{2\pi^c}{3}$	$\frac{3\pi^c}{4}$	$\frac{5\pi^c}{6}$	π^c

To calculate the sine, cosine and tangent of angles given in radians change the calculator mode to radians.

Example

Use your calculator to find the values of the following correct to 4 significant figures.

a) $\sin 0.52^c$ b) $\cos 1.3^c$ c) $\tan 1.14^c$

Answer

a) $\sin 0.52^c = 0.4969$

b) $\cos 1.3^c = 0.2675$

c) $\tan 1.14^c = 2.176$

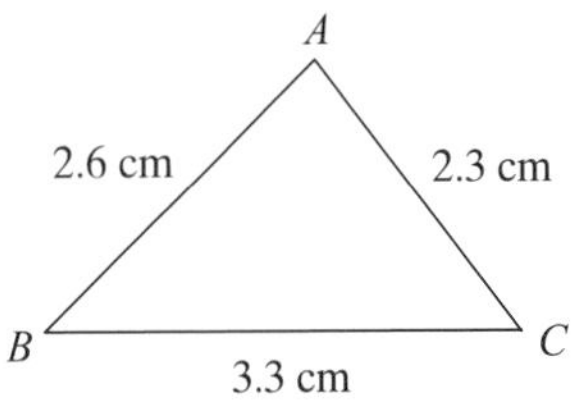

A, B, C; 2.6 cm, 2.3 cm, 3.3 cm

Fig. 4.18

Example

The three sides of a triangle have lengths 2.3 cm, 2.6 cm and 3.3 cm as shown in Figure 4.18. Find, in radians, the size of the largest angle.

Answer

The largest angle is opposite the longest side so we need to find the size of $\angle A$.

Cosine rule gives $\cos A = \frac{b^2 + c^2 - a^2}{2bc} = \frac{2.3^2 + 2.6^2 - 3.3^2}{2 \times 2.3 \times 2.6}$

$= 0.09699$

Using calculator in RAD mode gives $\angle A = 1.47^c$

Method notes

List the information given so $a = 3.3$, $b = 2.3$ and $c = 2.6$

We need to find $\angle A$ so use

$$\cos A = \frac{b^2 + c^2 - a^2}{2bc}$$

Arc length of a sector of a circle

Figure 4.19 shows a sector of a circle of radius r with an angle θ^c in the sector. The length of the arc of the sector is l.

The fraction of the circle occupied by the sector is given by the ratio of the length of the sector l to the circumference of the circle $2\pi r$ so the fraction occupied is $\frac{l}{2\pi r}$.

This fraction is also given by the ratio $\frac{\theta}{2\pi}$ (ratio of the angle in the sector to the angle in the full circle). Equating these ratios:

$$\frac{l}{2\pi r} = \frac{\theta}{2\pi} \text{ so } l = r\theta.$$

The length of the arc of a sector of radius r subtending an angle θ is $r\theta$ where θ is measured in radians.

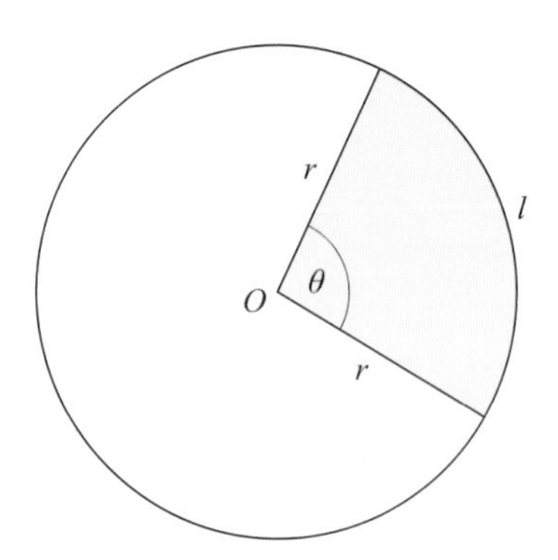

Fig. 4.19

Area of a sector of a circle

Referring again to Figure 4.19, the area of the circle is πr^2 so:

$$\frac{\text{area of sector}}{\pi r^2} = \frac{\theta}{2\pi} \text{ so area of sector} = \frac{1}{2}r^2\theta$$

The area of a sector of a circle of radius r subtending an angle θ is $\frac{1}{2}r^2\theta$ where θ is measured in radians.

Essential notes

π radians $\equiv 180°$

2π radians $\equiv 360°$

Example

A piece of wire of length 50 cm is bent to form a sector of a circle of radius 16 cm as shown in Figure 4.20.

a) Find the size of the angle in the sector.

b) Find the area of the sector.

Answer

a) **Step 1**: Perimeter of the sector is $2 \times 16 + l = 32 + l$

Step 2: If the wire is of length 50 cm then from step 1

$32 + l = 50$ so $l = 18$ cm.

Step 3: Arc length formula gives $l = r\theta$

$$\text{so } \theta = \frac{l}{r} = \frac{18}{16} = \frac{9}{8} = 1.125 \text{ radians}$$

b) Area of the sector $= \frac{1}{2}r^2\theta = \frac{1}{2} \times 16^2 \times 1.125 = 144 \text{ cm}^2$

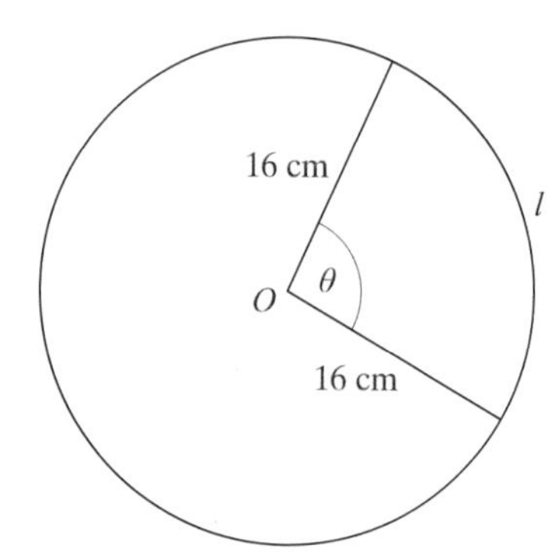

Fig. 4.20

Area of a segment in a circle

A segment of a circle is formed by a chord of the circle PQ and the arc length PQ (l) as shown in Figure 4.21. This segment (shaded yellow) has an area A.

A sector of a circle OPQ is drawn by joining P to O then O to Q then Q to P along the arc PQ (l) as shown in Figure 4.21.

$$\text{Area of the sector} = \frac{1}{2}r^2\theta$$

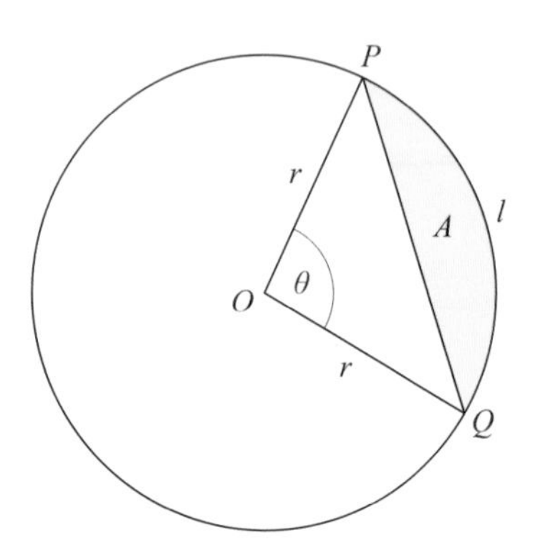

Fig. 4.21

Area of the triangle $OPQ = \frac{1}{2}pq \sin O$ where 'p' $= r$, 'q' $= r$ and '$\angle O$' $= \theta$

therefore area of $\Delta\ OPQ = \frac{1}{2}r^2 \sin\theta$

The area of the segment is the difference in the area of the sector of length l and the area of the triangle OPQ.

So the **area of the segment** in a circle is

$$A = \frac{1}{2}r^2\theta - \frac{1}{2}r^2 \sin\theta = \frac{1}{2}r^2(\theta - \sin\theta)$$

Example

A cylindrical pipe of diameter 1.8 metres contains water to a depth of 0.8 metres as shown in Figure 4.22. Find the cross sectional area of the water.

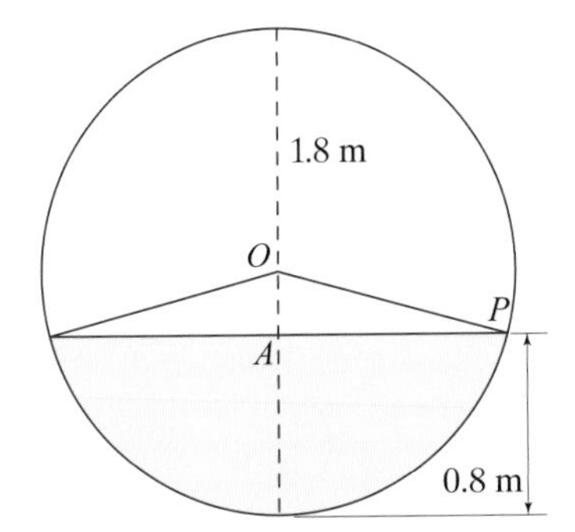

Fig. 4. 22
Cross section of a pipe with water of height 0.8 m.

Answer

Step 1: Diameter is 1.8 m so the radius of the sector is 0.9 m

Step 2: In $\Delta\ OAP$, $OP = 0.9$ m and $OA = 0.9 - 0.8 = 0.1$ m

so $\cos \angle POA = \frac{0.1}{0.9} = \frac{1}{9}$

so $\angle POA = 1.46^c$

Step 3: The angle of the sector $= 2 \times 1.46^c = 2.92^c$

Step 4: The area of the water (which forms a segment)

$= \frac{1}{2}r^2(\theta - \sin\theta)$

$= \frac{1}{2} \times 0.9^2 \times (2.92 - \sin 2.92)$

$= 1.09\ \text{m}^2$

Essential notes

Use RAD mode on your calculator throughout any question involving the sector of a circle.

Stop and think 2

If the pipe is of length 20 m find the volume of water in the pipe when the depth is 0.8 m.

Extending the trigonometric ratios

The trigonometric ratios for acute angles are listed below:

Fig. 4.23

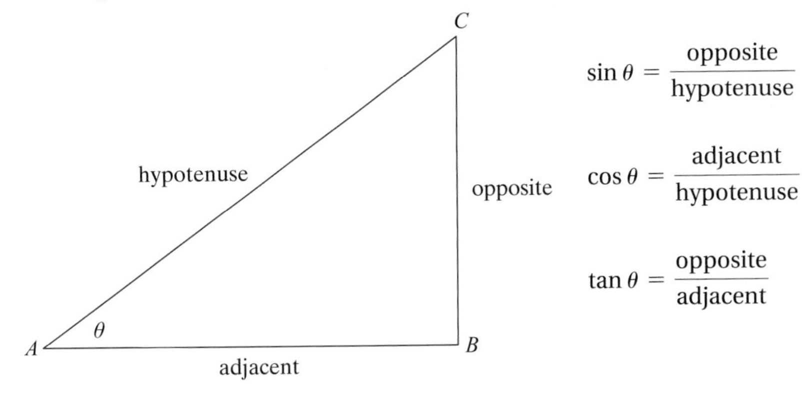

We now consider the trigonometric ratios for angles which are not acute.

By using the appropriate keys on a calculator in degrees mode we obtain the following results:

a) $\sin 120° = 0.8660$

b) $\cos 120° = -0.5$

c) $\tan 120° = -1.732$

d) $\sin 240° = -0.8660$

e) $\cos 240° = -0.5$

f) $\tan 240° = 1.732$

These answers show that for angles which are not acute the sine, cosine and tangent can be negative or positive depending on the size of the angle. To understand how this occurs consider 12 equally spaced points around a circle of radius 1 unit as shown in Figure 4.24. Each sector encloses an angle of 30°.

Essential notes

The three trigonometric ratios for acute angles were covered in the GCSE course.

An **acute angle** is an angle less than 90°.

An angle between 90° and 180° is called an **obtuse angle.**

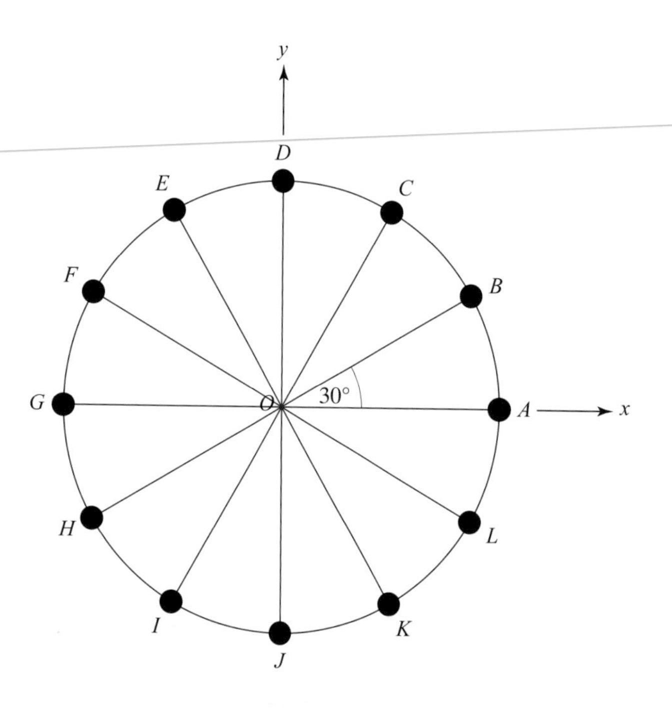

Fig. 4.24
A circle of radius 1 unit with 12 equally spaced points.

Method notes

12 equally spaced points and an angle of 360° at the centre means that each sector angle is $\frac{360°}{12} = 30°$.

Rotating the line OA anticlockwise through 120° (Figure 4.25) takes the line OA to OE so we know that $\angle EOS = 60°$ which is called the 'associated angle' for $\angle AOE$. The associated angle is always the **smallest** angle made with the x axis by the rotating line 'OA'.

Essential notes

Angles on a straight line add to give 180°.

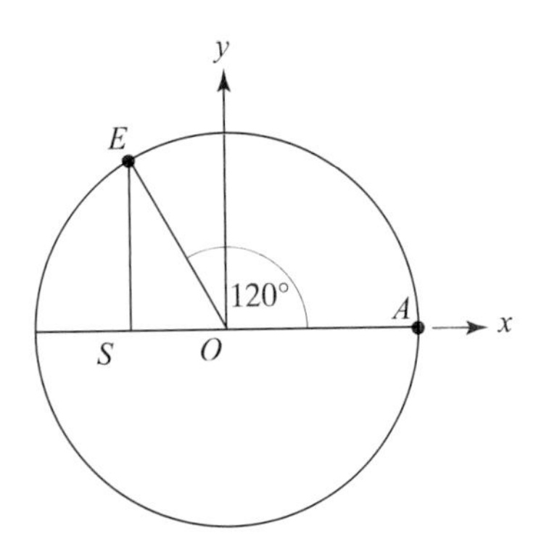

Fig. 4.25

To find cos 120° we use the 'associated' $\angle EOS$ in the right angled triangle *EOS*. We then take into account whether the lengths of the side of the triangle *OES* are in the positive or negative directions of the *x* and *y* axes. Conventionally, the rotating line '*OA*' is always considered to be in the positive direction wherever it 'stops' throughout a rotation so *OE* is positive. *OS* is in the negative direction of the *x* axis.

The angle for point *E* is 120°.

From the example we saw that:

$\sin 120° = 0.8660,$

$\cos 120° = -0.5$

$\tan 120° = -1.732.$

The *x* and *y* coordinates of *E* are $x = -0.5$ and $y = 0.8660$.

Consider triangle *OSE* in Figure 4.25:

$OS = -0.5$

$ES = 0.8660$

radius $OE = 1$

By considering the sign of the *x* and *y* coordinates we can extend the definitions of the trigonometric ratios to angles greater than 90° and less than 180°.

$$\sin 120^o = \frac{ES}{OE} = \frac{0.8660}{1} = 0.8660$$

$$\cos 120^o = \frac{OS}{OE} = \frac{-0.5}{1} = -0.5$$

$$\tan 120^o = \frac{ES}{OS} = \frac{0.8660}{-0.5} = -1.732$$

Essential notes

$\tan\theta = \dfrac{\sin\theta}{\cos\theta} = \dfrac{<0}{<0} > 0$ if $180° < \theta < 270°$.

These results will be applied in problem solving at the end of this section but first we must extend the trigonometric ratios for angles of any size.

Rotating the line *OA* anticlockwise though 240° takes the line *OA* to *OI* (Figure 4.26) so $\angle IOT = 60°$ which is the associated angle

Using $\Delta\ OIT$, $OI = 1$ so $\cos 60° = OT = -0.5$ (negative *x* direction) and $\sin 60° = IT = -0.8660$ (negative *y* direction)

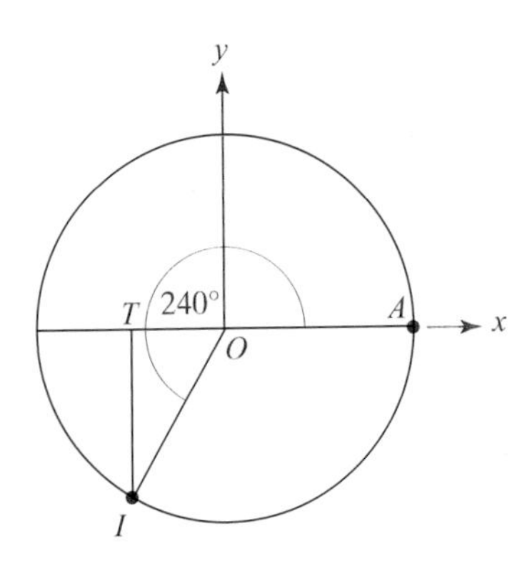

Fig. 4.26

By considering the sign of the *x* and *y* coordinates we can extend the definitions of the trigonometric ratios to angles greater than 180° and less than 270°.

$$\sin 240^o = \frac{IT}{OI} = \frac{-0.8660}{1} = -0.8660$$

$$\cos 240^o = \frac{OT}{OI} = \frac{-0.5}{1} = -0.5$$

$$\tan 240^o = \frac{OT}{IT} = \frac{-0.8660}{-0.5} = 1.732$$

Essential notes

$\tan\theta = \dfrac{\sin\theta}{\cos\theta} = \dfrac{<0}{<0} > 0$ if $180° < \theta < 270°$.

Rotating the line *OA* anticlockwise through 300° takes the line *OA* to *OK* (Figure 4.24 on page 61). $\angle AOK = 60°$ is the 'associated' angle since it is the

smallest angle made with the x axis by the rotating line OA. By using the same method as the x-coordinate of K is 0.5 and the y-coordinate of K is –0.8660 which gives:

$\sin 300° = -0.8660$

$\cos 300° = 0.5$

$\tan 300° = -1.732$

Essential notes

$\tan\theta = \frac{\sin\theta}{\cos\theta} = \frac{<0}{>0} < 0$ if $270° < \theta < 360°$.

Notation

Consider a point P moving round the circle as shown earlier in Figure 4.24.

Moving P round the circle in an **anti-clockwise** direction starting at A we pass from A to the points B, C, D, and so on in order.

The angle θ between the positive x-axis and OP increases from 0° at A to 30° at B, to 60° at C, to 90° at D, to 120° at E and so on. When the point P returns to A the line OP has gone through a full rotation of 360°. **Anti-clockwise angles are defined as positive.**

Moving P round the circle in a **clockwise** direction starting at A we pass through the points L, K, J, I, and so on in order.

In this case to describe what happens to the angle θ between the positive x axis and OP we use the convention that this angle decreases from 0° at A to –30° at L, to –60° at K, to –90° at J, to –120° at I and so on. When the point P returns to A the line OP has gone through a full rotation of –360°. **Clockwise angles are defined as negative.**

A mathematical convention is to divide the x-y plane into four quadrants beginning at the positive x axis and then to identify the signs of the trigonometric ratios in each quadrant.

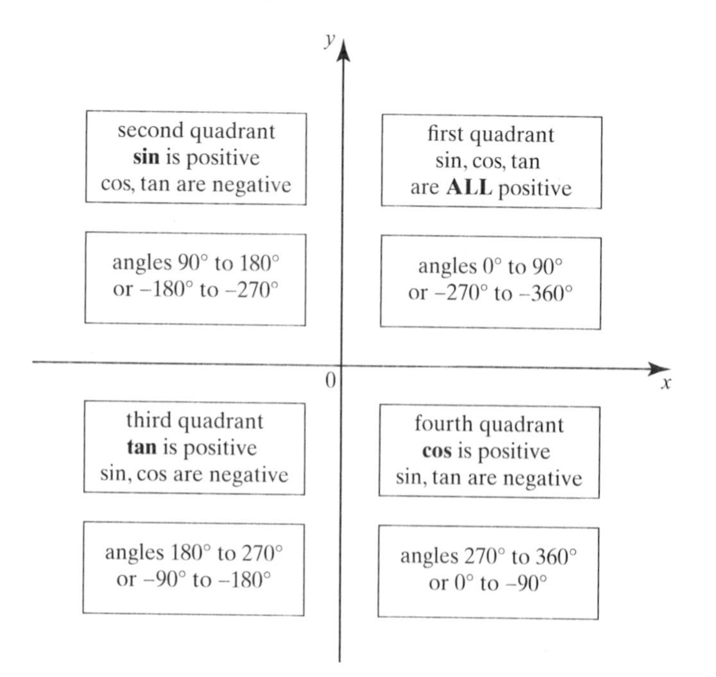

Essential notes

A quadrant is a quarter of a complete revolution.

Reading **anti-clockwise** from the bottom right quadrant and using the first letter of the trigonometric ratio which is **positive** as you move through each quadrant spells out **CAST.**

C cos fourth quadrant

A all first quadrant

S sin second quadrant

T tan – third quadrant

All the other ratios in each quadrant are negative.

The values for cosine and sine of any angle are best found using a calculator!

Sine, cosine and tangent of common angles

Example

Use the following diagrams to complete the table of trigonometric ratios for the common angles as listed below.

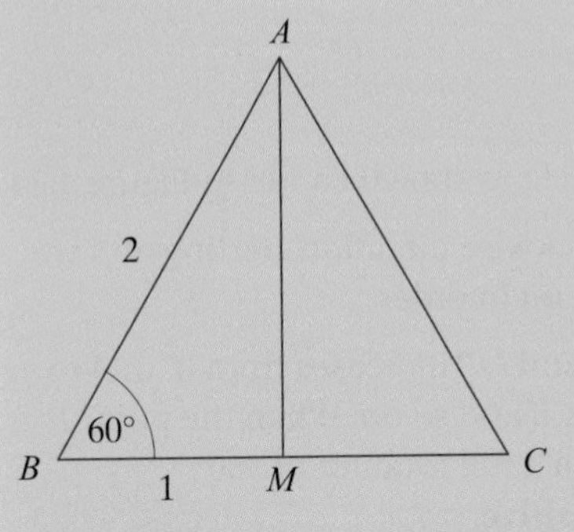

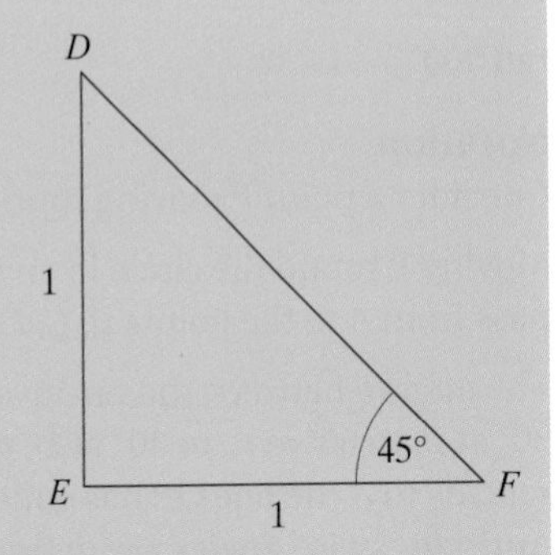

θ	0°	30°	45°	60°	90°
sin θ	0				1
cos θ	1				0
tan θ	0				∞

Answer

In triangle ABC, $AM = \sqrt{3}$ and angle $BAM = 30°$.

We can use this triangle for angles 30° and 60°.

In triangle DEF, $DF = \sqrt{2}$.

We can use this triangle for the angle 45°.

θ	0°	30°	45°	60°	90°
sin θ	0	$\frac{1}{2}$	$\frac{1}{\sqrt{2}} = \frac{\sqrt{2}}{2}$	$\frac{\sqrt{3}}{2}$	1
cos θ	1	$\frac{\sqrt{3}}{2}$	$\frac{1}{\sqrt{2}} = \frac{\sqrt{2}}{2}$	$\frac{1}{2}$	0
tan θ	0	$\frac{1}{\sqrt{3}}$	1	$\sqrt{3}$	∞

Method notes

In $\triangle ABM$ using Pythagoras' Theorem

$2^2 = 1^2 + AM^2$

$4 = 1 + AM^2$

$3 = AM^2$ so $AM = \sqrt{3}$

In $\triangle DEF$ using Pythagoras' Theorem

$DF^2 = 1^2 + 1^2$

$DF^2 = 1 + 1$

$DF^2 = 2$ so $DF = \sqrt{2}$

Essential notes

It is essential to remember the values for 0° and 90°.

It is also useful to remember the sines and cosines for angles 30°, 45° and 60° in surd form or how to derive them quickly from a triangle.

Stop and think 2

Without using a calculator work out the following, giving reasons for your answers:

a) sin 135° b) cos (–120°) c) tan (–270°)

and give your answers in surd form where appropriate.

Graphs of trigonometric functions

You have seen that the definitions of cos θ and sin θ are derived from the x and y coordinates of a point P moving round the circumference of a circle of radius 1.

Using a calculator for the angles listed below gives the values for cosine and sine.

The answer for each of the angles listed below is found by using a calculator in **degrees** mode. Be careful to read the **full** answer from you calculator as it will indicate if it is a negative answer.

cos 0° = 1	sin 0° = 0
cos 30° = 0.8660	sin 30° = 0.5
cos 45° = 0.7071	sin 45° = 0.7071
cos 60° = 0.5	sin 60° = 0.8660
cos 90° = 0	sin 90° = 1
cos 120° = −0.5	sin 120° = 0.8660
cos 135° = −0.7071	sin 135° = 0.7071
cos 150° = −0.8660	sin 150° = 0.5
cos 180° = −1	sin 180° = 0
cos 210° = −0.8660	sin 210° = −0.5
cos 225° = −0.7071	sin 225° = −0.7071
cos 240° = −0.5	sin 240° = −0.8660
cos 270° = 0	sin 270° = −1
cos 300° = 0.5	sin 300° = −0.8660
cos 315° = 0.7071	sin 315° = −0.7071
cos 330° = 0.8660	sin 330° = −0.5
cos 360° = 1	sin 360° = 0

From these values we can sketch a graph of the **trigonometric functions** $y = \cos\theta$ and $y = \sin\theta$ as shown in Figures 4.28 and 4.29.

Because the values of y are derived from a point moving round the circumference of a circle $y = \cos\theta$ and $y = \sin\theta$ are called **circular functions**.

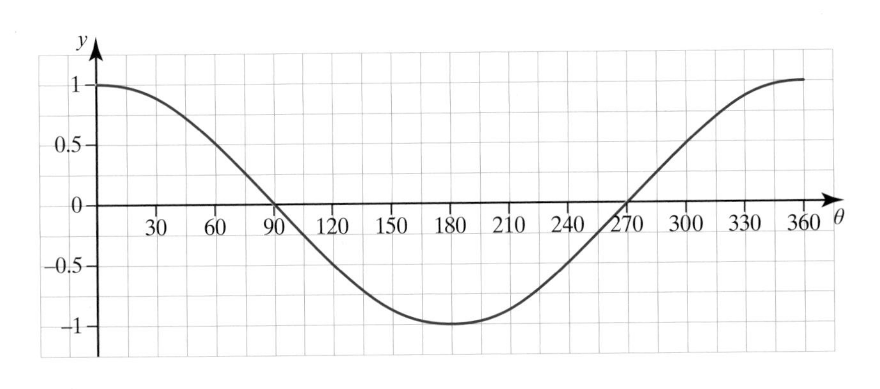

Fig. 4.28
Graph of $y = \cos\theta$

Fig. 4.29
Graph of $y = \sin\theta$

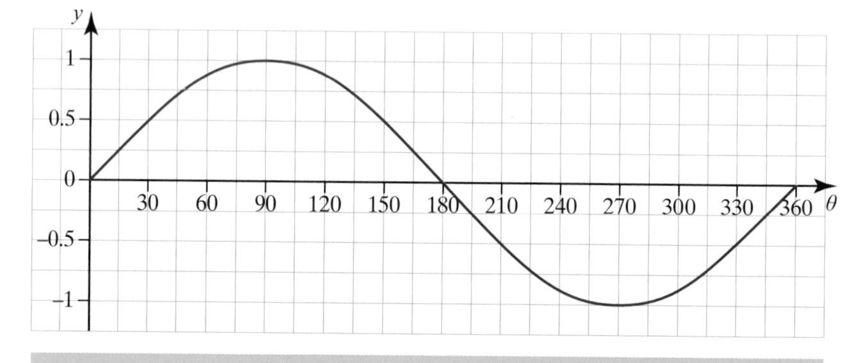

Method notes

Choose any value for y so that you can read from the cosine graph the corresponding value for θ.

An easy value to choose is $y = 0.5$.

Draw the line $y = 0.5$ on the cosine graph and read off the value of θ where the line crosses the curve which is $\theta = 60°$ and $300°$

Repeat this with different values for y on each graph.

Example

Use the graphs shown in Figure 4.28 and Figure 4.29 to verify that

$\cos\theta = \cos(360° - \theta)$

$\sin\theta = -\sin(360° - \theta)$ $0° \leq \theta \leq 360°$

$\sin\theta = \sin(180° - \theta)$

$\cos\theta = -\cos(180° - \theta)$ $0° \leq \theta \leq 180°$

Answer

From the cosine graph we see that $\cos 60° = \cos 300°$ and $\cos 210° = \cos 150°$ as examples of the rule $\cos\theta = \cos(360° - \theta)$

From the sine graph we see that $\sin 60° = -\sin 300°$ and $\sin 210° = -\sin 150°$ as examples of the rule $\sin\theta = -\sin(360° - \theta)$

From the sine graph we see that $\sin 30° = \sin 150°$ is an example of the rule $\sin\theta = \sin(180° - \theta)$.

From the cosine graph we see that $\cos 30° = -\cos 150°$ as an example of the rule $\cos\theta = -\cos(180° - \theta)$.

Essential notes

You saw this idea earlier in the Chapter when using the sine rule to solve for side lengths in triangles.

This example shows that for a chosen value of y there is more than one solution to a equation of the form $y = \sin\theta$ or $y = \cos\theta$. This is explored further in the following example.

Method notes

For each equation:

The first (or primary) solution comes from a calculator.

The second solution can be found by using the CAST diagram.

Example

Find all the solutions of the following equations for $0° < \theta < 360°$

a) $\sin\theta = 0.7$ b) $\cos\theta = -0.4$

Answer

a) Using a calculator: $\sin\theta = 0.7$ so $\theta = 44.43°$

The CAST diagram shows that sine is positive in the second quadrant so another solution is $180° - 44.43° = 135.57°$

b) Using a calculator: $\cos\theta = -0.4$ so $\theta = 113.58°$

The CAST diagram shows that cosine is negative in the third quadrant so another solution is $360° - 113.58° = 246.42°$

Sine and cosine waves

Figures 4.28 and 4.29 show the basic shapes of the sine and cosine graphs. Because the values of x and y can be obtained from the coordinates of a point moving round a circle of radius 1, the graphs can be extended beyond the range 0° to 360° as the point can move through more than one revolution in the positive or negative direction.

We now adopt our conventional function notation of writing y as a function of x. Figures 4.30 and 4.31 show the graphs of $y = \cos x$ and $y = \sin x$ for x in the interval $-360° \leq x \leq 360°$.

The size of the interval does not have to be restricted to the interval $-360° \leq x \leq 360°$. The wave shape of the graphs may be extended indefinitely for any angle, positive or negative.

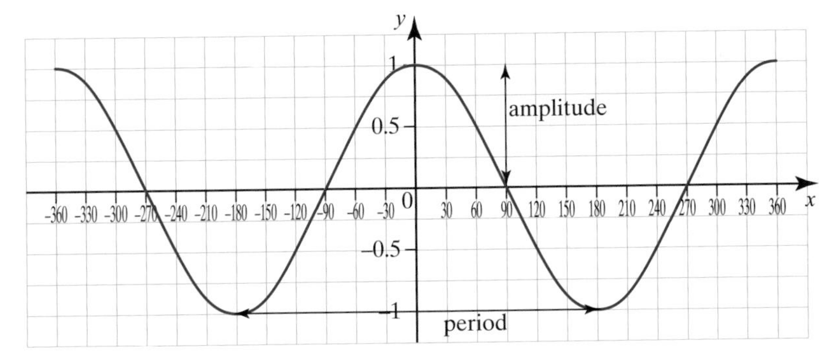

Fig. 4.30
Graph of $y = \cos x$.

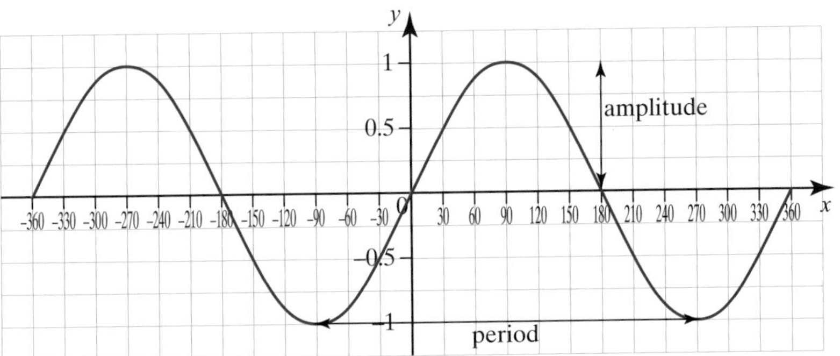

Fig. 4.31
Graph of $y = \sin x$.

The wave forms repeat themselves every 360°. This is called the **period** of the wave. Any function with a graph which exhibits a repeating pattern is called **periodic**. The waves oscillate between −1 and +1. We say that the wave has **amplitude** of 1 unit. The sine wave is often used for problem solving in physics.

From Figure 4.30 notice that $\cos(-x) = \cos(x)$ e.g $\cos(-60) = \cos(60)$. The cosine curve has reflectional symmetry in the y-axis. We say that cosine is an **even function**.

From Figure 4.31 notice that $\sin(-x) = -\sin(x)$ e.g $\sin(-60) = -\sin(60)$. The sine curve has rotational symmetry about the origin. We say that sine is an **odd function**.

Method notes

$\sin = \dfrac{\text{opposite}}{\text{hypotenuse}}$

$\cos = \dfrac{\text{adjacent}}{\text{hypotenuse}}$

$\dfrac{\sin}{\cos} = \dfrac{\text{opp}}{\text{hyp}} \div \dfrac{\text{adj}}{\text{hyp}}$

$\dfrac{\sin}{\cos} = \dfrac{\text{opp}}{\text{adj}} = \tan$

Graph of the tangent function

Figure 4.28 shows the graph of $y = \tan x$ for x in the interval $-360° \leq x \leq 360°$.

Values of the tangent function can be calculated from $y = \tan x = \dfrac{\sin x}{\cos x}$.

The tangent graph does not form a wave but it is periodic with period 180°. It is an **odd** function like the sine graph. The range of values for the tangent graph is infinite.

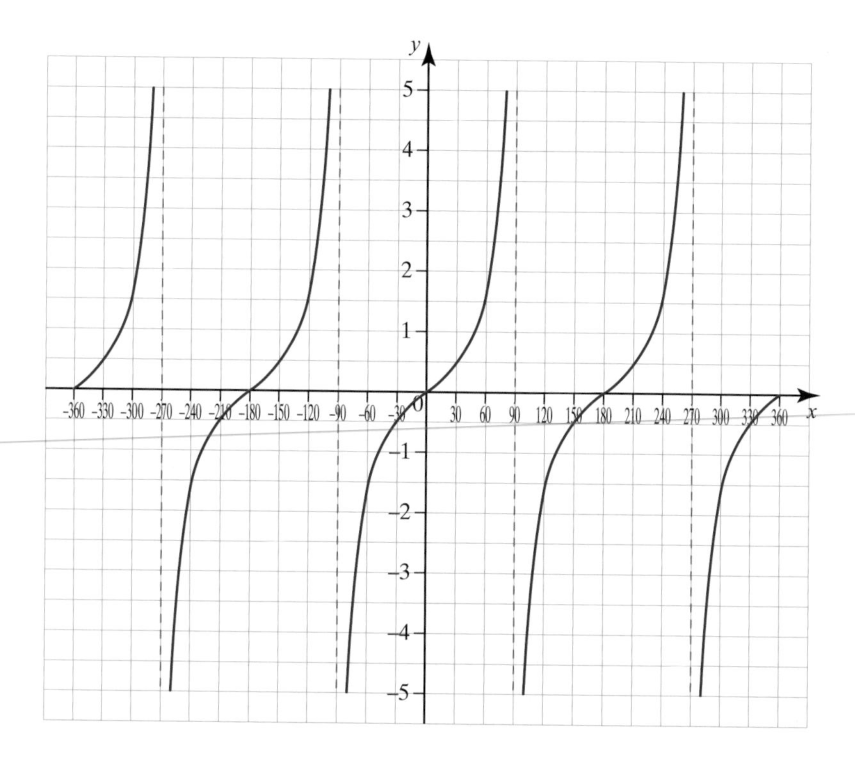

Fig. 4.32
Graph of $y = \tan x$

The dashed lines are called **asymptotes**. These occur at the values of x where $\tan x$ has infinite value e.g. $x = -270°$, $x = -90°$, $x = 90°$, $x = 270°$.

Trigonometric identities

An important identity known as the **Pythagorean identity** is useful for the manipulation of trigonometric expressions.

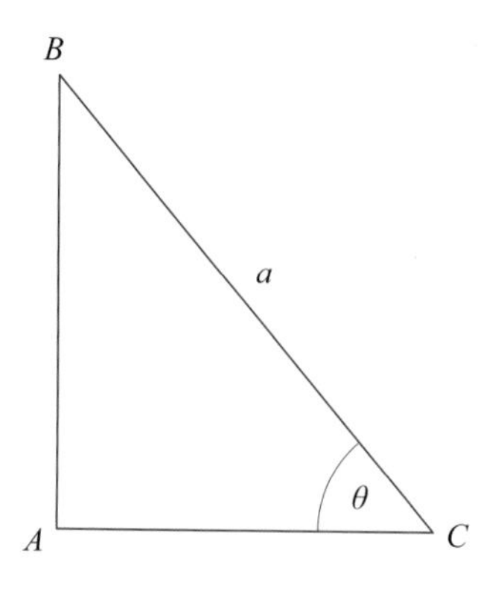

Fig. 4.33

Figure 4.33 shows a right angled triangle with hypotenuse of length a and base angle θ:

$$BC = a \cos \theta$$

$$AB = a \sin \theta$$

Apply Pythagoras' Theorem to the triangle ABC:

$$AB^2 + BC^2 = AC^2$$

$$\Rightarrow (a \sin \theta)^2 + (a \cos \theta)^2 = a^2$$

$$\Rightarrow a^2 \sin^2 \theta + a^2 \cos^2 \theta = a^2$$

Dividing each term by a^2 gives

$\sin^2\theta + \cos^2\theta \equiv 1$

This is an example of an identity which means it is true for all values of θ not just a restricted number of values of θ and the identity symbol is $\equiv$.

Example

Use the Pythagorean identity to find the exact value of $\cos\alpha$ given that $\sin\alpha = \frac{3}{5}$ and that α is acute.

Answer

$\sin^2\alpha + \cos^2\alpha \equiv 1$

$\Rightarrow \cos^2\alpha = 1 - \sin^2\alpha$

$\Rightarrow \cos^2\alpha = 1 - \left(\frac{3}{5}\right)^2 = \frac{25-9}{25} = \frac{16}{25}$

$\Rightarrow \cos\alpha = \pm\frac{4}{5}$

Since α is acute $\cos\alpha = \frac{4}{5}$

Acute angles are first quadrant angles therefore cosine is positive.

Exam tips

Learn the two identities:

$\frac{\sin x}{\cos x} \equiv \tan x$

$\sin^2\theta + \cos^2\theta \equiv 1$

where x and θ are general angles.

Example

Write the following equations as equations involving only one trigonometric function

a) $\cos^2\theta - \sin^2\theta = 0$

b) $\sin x + \cos x = 3\sin x - 2\cos x$

Answer

a) $\cos^2\theta - \sin^2\theta = 0$

Step 1: Rearranging the Pythagorean identity

$\sin^2\theta + \cos^2\theta \equiv 1$

$\Rightarrow \sin^2\theta \equiv 1 - \cos^2\theta$

Step 2: Substitute from step 1 into:

$\cos^2\theta - \sin^2\theta = 0$

$\Rightarrow \cos^2\theta - (1 - \cos^2\theta) = 0$

$\Rightarrow 2\cos^2\theta - 1 = 0$

which gives the original equation in terms of $\cos\theta$ only.

Continued on next page

Method notes

The Pythagorean identity

$\sin^2\theta + \cos^2\theta \equiv 1$

can be rearranged to give:

$\sin^2\theta \equiv 1 - \cos^2\theta$ or

$\cos^2\theta \equiv 1 - \sin^2\theta$

Always choose the most suitable rearrangement to answer the question given. In (a) you could choose either rearrangement.

b) $\sin x + \cos x = 3\sin x - 2\cos x$

Step 1: Collect like terms:

$$3\cos x = 2\sin x$$

Step 2: Divide both sides by $\cos x$:

$$3 = \frac{2\sin x}{\cos x}$$ (we have assumed $\cos x$ is not equal to 0)

Step 3: Use the identity that $\frac{\sin x}{\cos x} \equiv \tan x$ in step 2:

$$3 = 2\tan x$$

$$\Rightarrow \tan x = \frac{3}{2}$$

Method notes

You must be able to state $\sin^2 x + \cos^2 x \equiv 1$ for any angle x.

Example

Show that $(\sin x + \cos x)^2 = 1 + 2\sin x\cos x$

Answer

Use the left hand side of the given equation: this side has been chosen as it is often easier to multiply out terms and then simplify them rather than vice versa.

Step 1: $(\sin x + \cos x)^2 = \sin^2 x + \cos^2 x + 2\sin x\cos x$

Step 2: Use the Pythagorean identity in step 1 to give $1 + 2\sin x\cos x$ which equals the right hand side of the original equation as required.

Method notes

These examples illustrate **direct proof** or **algebraic proof** in mathematics.
In such proofs you must start with the information given on **one side** of the original statement and by algebraic manipulation of this information you must deduce the information on the **other side** of the original statement!

Example

Show that $\sin^2 x\cos^2 y - \cos^2 x\sin^2 y = \sin^2 x - \sin^2 y$

Answer

We see from the right hand side of the given equation that it is written only in terms of the sine ratio. Therefore in using the left hand side we must rewrite cosines in terms of sines.

Step 1: Using the Pythagorean identity $\cos^2 x \equiv 1 - \sin^2 x$ in the left hand side of the original equation:

$$\sin^2 x\cos^2 y - \cos^2 x\sin^2 y = \sin^2 x\cos^2 y - (1 - \sin^2 x)\sin^2 y$$

Step 2: Simplify: $\sin^2 x\cos^2 y - \sin^2 y + \sin^2 x\sin^2 y$

Step 3: Factorise: $\sin^2 x(\cos^2 y + \sin^2 y) - \sin^2 y$

Step 4: Use the Pythagorean identity in step 3 to obtain $\sin^2 x - \sin^2 y$ as required.

Trigonometric equations

Solving simple equations involving sine, cosine and tangent for angles between 0° and 90°. was covered earlier in the course. Now we must solve more complicated equations, often involving more than one solution and two or more trigonometric functions.

Example

Solve the equation $4\cos\theta - 3\sin\theta = 0$, $0° \le \theta \le 360°$.

Answer

We need to rewrite $4\cos\theta - 3\sin\theta = 0$ as a statement involving one trigonometric function before we can sketch a graph.

Step 1: Rewrite $4\cos\theta - 3\sin\theta = 0$ so $4\cos\theta = 3\sin\theta$

Step 2: Simplify step 1 to give $\dfrac{\sin\theta}{\cos\theta} = \dfrac{4}{3}$ so $\tan\theta = \dfrac{4}{3}$

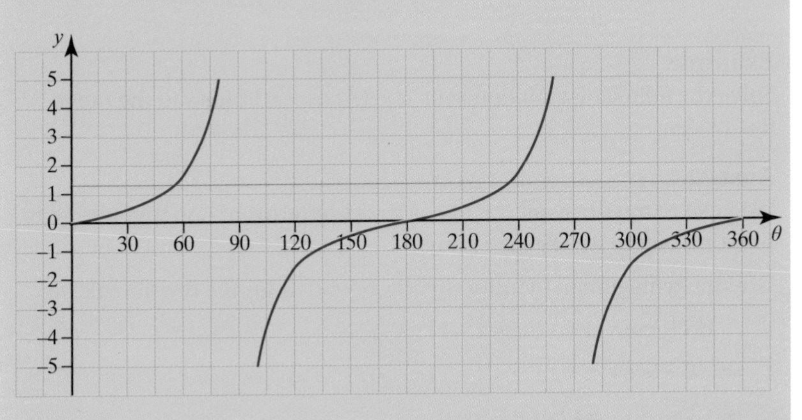

θ = the angle which has a tangent $= \dfrac{4}{3}$ written as $\tan^{-1}\dfrac{4}{3} = 53.13°$ (from the calculator)

$\theta = 180° + 53.13° = 233.13°$ (from the graph)

Method notes

Sketch a graph of $y = \tan\theta$ in the given range for θ. Draw in the line $y = \dfrac{4}{3}$ to identify how many solutions there are (where line and curve cross). Here there are 2 solutions. The calculator gives the **primary solution** and the graph helps to find the **second solution.**

Fig 4.34
The solutions of $\tan\theta = \dfrac{4}{3}$

Fig. 4.35
The solutions of $\sin x = \dfrac{1}{3}$ and $\sin x = -\dfrac{1}{2}$

Example

Solve the equation $6\sin^2 x + \sin x - 1 = 0$ for x in the range $-180° \le x \le 180°$.

Answer

$6\sin^2 x + \sin x \equiv 1 = (3\sin x - 1)(2\sin x + 1) = 0$

$3\sin x - 1 = 0$ so $\sin x = \dfrac{1}{3}$, $2\sin x + 1 = 0$ so $\sin x = -\dfrac{1}{2}$

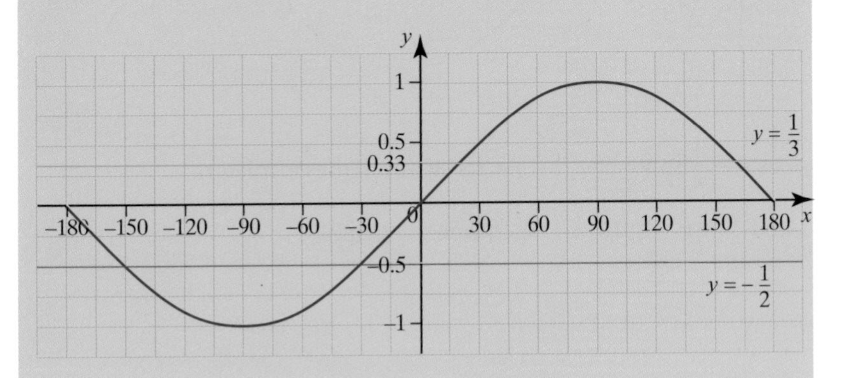

Method notes

Factorise the quadratic in $\sin x$ and solve to give two values for $\sin x$.

We must now sketch a graph of $y = \sin x$ **in the given range for** x then draw in the lines $y = \dfrac{1}{3}$ and $y = -\dfrac{1}{2}$ to identify the number of solutions. Here there are four solutions. The calculator gives the **primary solution** and the graph helps to find the second solution for each value of $\sin x$.

$\sin x = \frac{1}{3}$ so $x = \sin^{-1}\frac{1}{3} = 19.47°$ (from the calculator)

$x = 180° - 19.47° = 160.53°$ (from the graph)

$\sin x = -\frac{1}{2}$ so $x = \sin^{-1}\left(-\frac{1}{2}\right) = -30°$ (from the calculator)

$x = -180° - (-30°) = -150°$ (from the graph).

You may be asked to give the answers to some questions in radians instead of degrees.

Essential notes

An angle range given in terms of π means that **radians** are being used to measure the angles not degrees.

$\pi^c \equiv 180°$

$2\pi^c \equiv 360°$

Example

Solve the following equation $8\sin^2 x + 14\cos x = 11$ for x in the range $0 \leq x \leq 2\pi$.

Answer

We need to rewrite $8\sin^2 x + 14\cos x = 11$ in terms of one trigonometric function before we sketch a graph.

Use the Pythagorean identity $\sin^2 x = 1 - \cos^2 x$ for the rewrite to give:

$$8(1 - \cos^2 x) + 14\cos x = 11$$

$$\Rightarrow 8 - 8\cos^2 x + 14\cos x = 11$$

$$\Rightarrow 8\cos^2 x - 14\cos x + 3 = 0$$

$$\Rightarrow (4\cos x - 1)(2\cos x - 3) = 0$$

$$4\cos x - 1 = 0 \Rightarrow \cos x = \frac{1}{4}$$

$$2\cos x - 3 = 0 \Rightarrow \cos x = \frac{3}{2}$$

$\cos x$ has a maximum value of 1 therefore $\cos x = \frac{3}{2}$ gives no solutions.

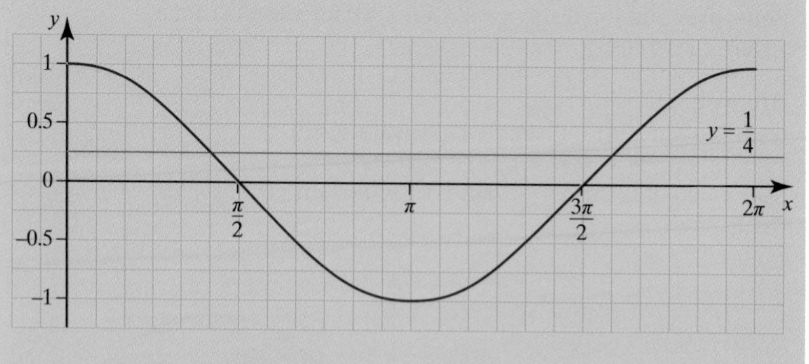

Fig. 4.36
The solutions of $\cos x = \frac{1}{4}$

$\cos x = \frac{1}{4}$ so $x = \cos^{-1}\left(\frac{1}{4}\right) = 1.318^c$ (from the calculator)

and $x = 2\pi - 1.318^c = 4.965^c$ (from the graph)

Method notes

The range for x is given in radians: remember to check that your calculator is in **RAD (radian)** mode throughout this question.

Solve the quadratic equation in $\cos x$ to give the (apparent) solutions. In the sketch of the graph of $y = \cos x$ use radians on the x axis.

Draw in the line $y = \frac{1}{4}$ which crosses the graph $y = \cos x$ twice.

The calculator gives the **primary** solution. The graph helps to find the second solution.

Stop and think answers

1

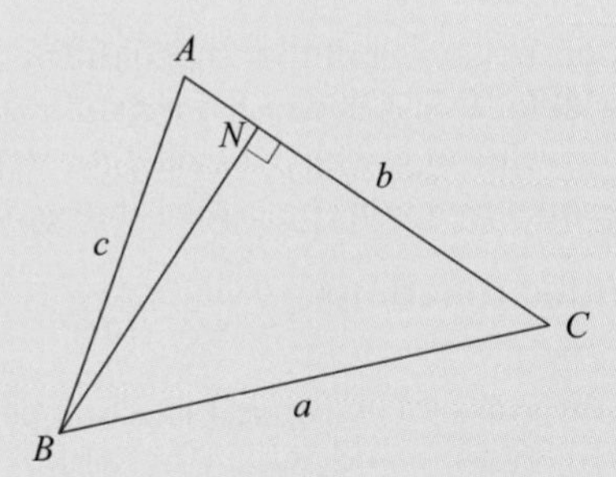

Fig. 4.37

In Δ ABN: $\sin A = \dfrac{BN}{c}$ so $BN = c \sin A$ (1)

Area of Δ ABC $= \dfrac{1}{2} \times$ base $\times$ height $= \dfrac{1}{2} \times b \times BN$

so substituting for BN from equation (1) gives area of:

$\Delta ABC = \dfrac{1}{2} bc \sin A$

2. Area of cross section = 1.09 m^2 when the water has a depth of 0.8 m.

   If the length of the pipe is 20 m then the volume of water in the pipe is:

   area of cross section × length = 1.09 × 20 = 21.8 m^3

3. Without using a calculator we can find the answers using common angles:

   $\sin 135° = \sin (180 - 135)° = \sin 45° = \dfrac{1}{\sqrt{2}}$

   Taking the positive sense (anticlockwise) for measuring angle to get to the same point gives $\cos (-120)° = \cos 240° = -\cos 60°$

   This is a third quadrant angle so cos is negative therefore:

   $\cos (-120°) = -0.5.$

   $\tan (-270)° = \tan 90°$ (same method as in (b)) $= \infty$

   $\Rightarrow \tan (-270)° = \infty$

Exponential functions $y = a^x$

Exponential functions are functions of the form $y = a^x$, where a is a constant.

The constant a is a non-negative number called the **base** and x is called the **independent variable**. This is because the x value given in the function $y = a^x$ determines the value of y, which is called the **dependent variable**. The graphs of exponential functions have the same general shape.

Graphs of exponential functions

Key points

The general form of the graph of an exponential function is similar to the graph of $y = 2^x$ shown below.

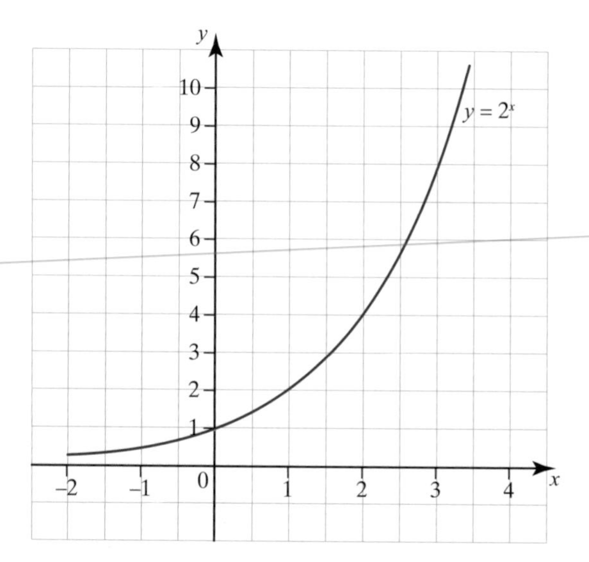

Fig. 5.1
Graph of the exponential function $y = 2^x$.

The key features to note in a graph of this exponential function are:

- As $x \to \infty$ then $y \to \infty$: this means that for positive values of x, as x gets larger and larger the y gets larger and larger.
- As $x \to -\infty$ then $y \to 0$: this means that as x gets larger and negative the graph approaches the x axis but never reaches it.
- The graph passes through (0, 1) because $a^0 = 1$ for any number a (this was covered in Core 1)
- The value of y is always positive.
- The gradient of a curve gets larger (at an increasing rate) as x gets larger and it is always positive

Figure 5.2 shows graphs of the exponential functions $y = 3^x$, $y = 2^x$ and $y = 1.5^x$.

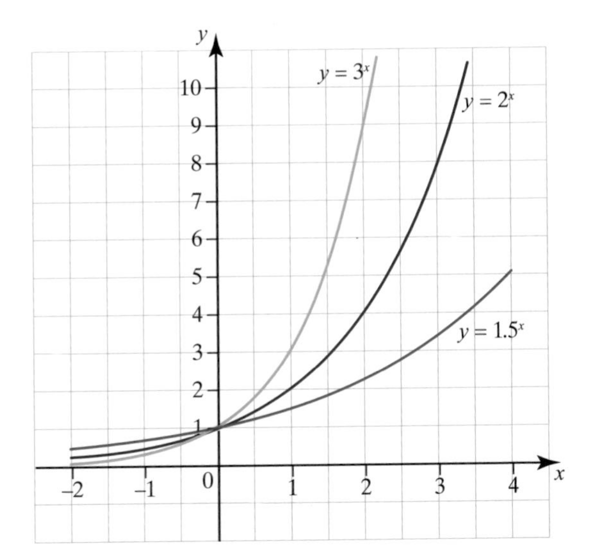

Fig. 5.2
Graphs of the exponential functions $y = 3^x$, $y = 2^x$ and $y = 1.5^x$.

Method notes

The graphs are of $y = a^x$, $a^0 = 1$ for all values of x because if $x = 0$ $y = 1$

Evaluating the gradients of the tangents to the curves for different values of x gives the **general results** shown.

Note that each graph passes through the point (0, 1). The gradient functions of exponential graphs are similar to the functions themselves as shown in these **general results**:

for $y = 2^x$ the gradient function is $\frac{dy}{dx} = 0.69 \times 2^x$;

for $y = 3^x$ the gradient function is $\frac{dy}{dx} = 1.1 \times 3^x$

for $y = 1.5^x$ the gradient function is $\frac{dy}{dx} = 0.41 \times 1.5^x$

Essential notes

The numbers 0.69, 1.1 and 0.41 are given correct to 2 significant figures. You will not need to remember or prove these results.

Figure 5.3 shows graphs of the exponential functions $y = 2^x$ and $y = (\frac{1}{2})^x$. The function $y = (\frac{1}{2})^x$ can also be written as $y = 2^{-x}$. The value of $y = 2^{-x}$ is always positive and the gradient of the graph is always negative.

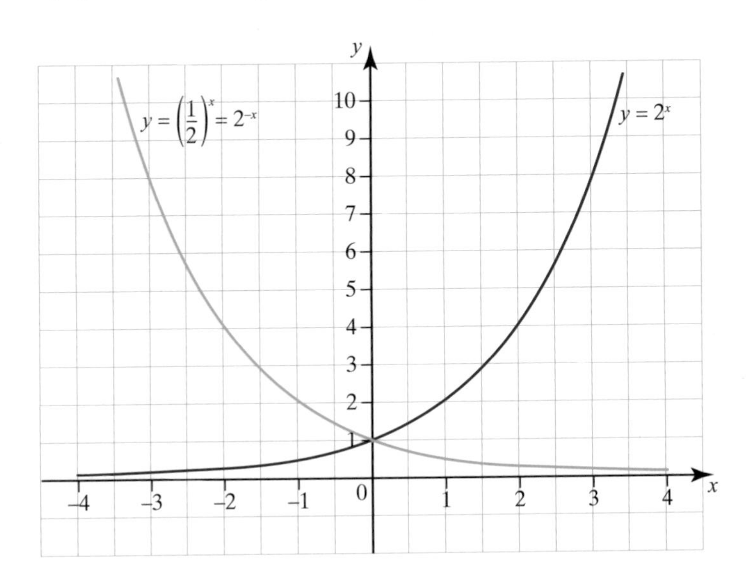

Fig. 5.3
Graphs of the exponential functions $y = 2^x$ and $y = (\frac{1}{2})^x$.

Exam notes

You must learn the shape of the graph of the functions $y = a^x$. When $a > 1$ the graph is increasing as the slope of the tangent at each point is positive. When $0 < a < 1$ the graph is decreasing as the slope of the tangent at each point is negative.

Modelling with exponential functions

Method notes

The initial population is 2 (million) and growing at 2% per year. This means at the end of the first year the population is $2 + \frac{2}{100} \times 2 = 2(1 + 0.02) = 2(1.02)$.

At end of the second year it is $2(1.02)(1.02) = 2(1.02)^2$.

At end of t years it is $2(1.02)^t$.

1.02^t is mathematically the same as compound interest of 2% which was discussed in Core 1.

Example

The population of a country, currently 2 million, is growing at a rate of two per cent per annum.

a) Show that the expected population, p millions, in t years time, is given by $p = 2 \times 1.02^t$

b) Sketch a graph of p against t for $0 \leq t \leq 100$

c) Use your graph to estimate:

i) the size of the population in 35 years' time,

ii) the time taken for p to reach 10 million.

Answer

a) If the population is growing at 2 per cent per year, then each year the population is 1.02 times larger than it was at the start of the year before. Thus after t years:

$$p = 2 \times 1.02^t$$

b) A sketch of p against t is shown in Figure 5.4 below.

Fig. 5.4

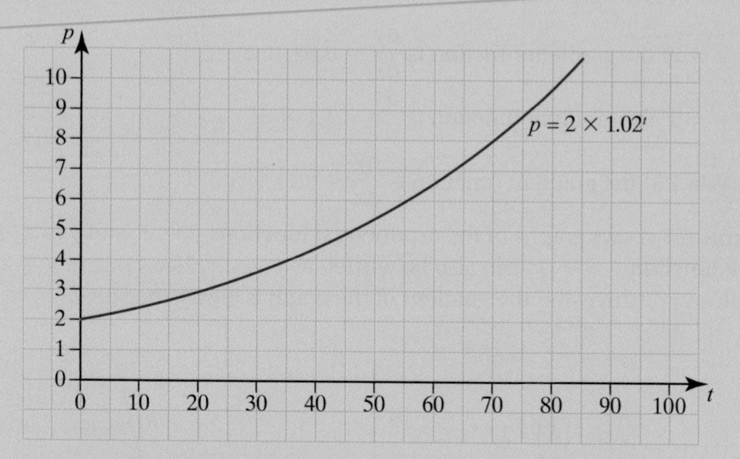

c) i) From the graph, after 35 years the corresponding value for p is approximately 4 million.

ii) From the graph, the population is 10 million in just over 80 years time.

Powers of 10

A useful exponential function which appears on scientific calculators is 10^x. You will be familiar with the common use of the function 10^x for expressing very large numbers and very small numbers. This is called **standard form**. For example, the mass of the Earth is 5 980 000 000 000 000 000 000 000 kg which is written in standard form as 5.98×10^{24}, a hydrogen atom has mass 0.000 000 000 000 000 000 000 001 67 grams which is written in standard form as 1.67×10^{-24}.

In general, a number written in standard form is expressed as $a \times 10^n$ where $1 < a < 10$ and n is an integer.

Essential notes

Standard form was covered in GCSE courses. An integer is a whole number.

Example

Find the value of y for each of the given values of x.

a) $y = 3^x$ at $x = 2$

b) $y = 10^x$ at $x = 4$

c) $y = 2.7^x$ at $x = 3$

d) $y = \frac{1^x}{2}$ at $x = 1.4$

e) $y = 10^x$ at $x = 2.7$

Answer

a) $y = 3^x$ at $x = 2$ so $y = 3^2 = 9$

b) $y = 10^x$ at $x = 4$ so $y = 10^4 = 10000$

c) $y = 2.7^x$ at $x = 3$ so $y = 2.7^3 = 19.683$

d) $y = \left(\frac{1}{2}\right)^x$ at $x = 1.4$ so $y = \left(\frac{1}{2}\right)^{1.4} = 0.37893$

e) $y = 10^x$ at $x = 2.7$ so $y = 10^{2.7} = 501.187$

Method notes

These calculations are easily carried out on a scientific calculator.

Use the 'power' key (often shown as Λ) on your calculator.

If we are given the value of the base of an exponential function $y = a^x$ then we can find y-values for given x-values quite easily either by mental methods as shown in (a) and (b) above or at the press of a key on a calculator as in (c), (d) and (e)!

However, if you are given y values, finding the x values is often not quite so straightforward. Much will depend on the actual y values given as is illustrated in the following examples.

Essential notes

If $y = a^x$ a is the base.

Example

Find the value of x for each of the given values of y.

a) $y = 3^x$ at $y = 9$

b) $y = 10^x$ at $y = 1000$

c) $y = 2^x$ at $y = 32$

Answer

a) $y = 3^x$ at $y = 9$ so $3^x = 9 = 3^2$ so $x = 2$

b) $y = 10^x$ at $y = 1000$ so $10^x = 1000 = 10^3$ so $x = 3$

c) $y = 2^x$ at $y = 32$ so $2^x = 32 = 2^5$ so $x = 5$

Method notes

To solve $10^x = 4.6$ we could use a method of trial and improvement:

$x = 0$ so $10^0 = 1$

$x = 1$ so 10^1

which shows x is between 0 and 1. Then using a calculator and taking various values of x with $0 < x < 1$ so if $10^x = 4.6$ we find $x = 0.7$ to 1 decimal place. However, a faster, accurate method is to use the logarithmic function.

In the examples above, the answers were quite easy to find because we could write each number as a power of the base e.g. it was easy to write $1000 = 10^3$.

But consider the problem of finding x when $y = 10^x$ at $y = 4.6$. It is not so easy this time to find the power x such that $10^x = 4.6$

Consider the two equivalent statements: if $y = x^2$ then $x = \pm\sqrt{y}$. The square root function 'undoes' squaring a number. For example, find the value of x if $x^2 = 6.3$. Then $x = \pm\sqrt{6.3} = \pm 2.51$

We need a function for 'undoing' the exponential function, i.e. we need an equivalent statement for $y = a^x$.

This function is called the logarithmic function.

Logarithmic functions

Definition: If $a^x = y$ then $x = \log_a y$ which we read as:

'x is the logarithm to base a of y'

$a^x = y$ and $x = \log_a y$ are equivalent statements.

Figure 5.5 below summarizes the relationship between an exponential function and its associated logarithm function.

Fig. 5.5

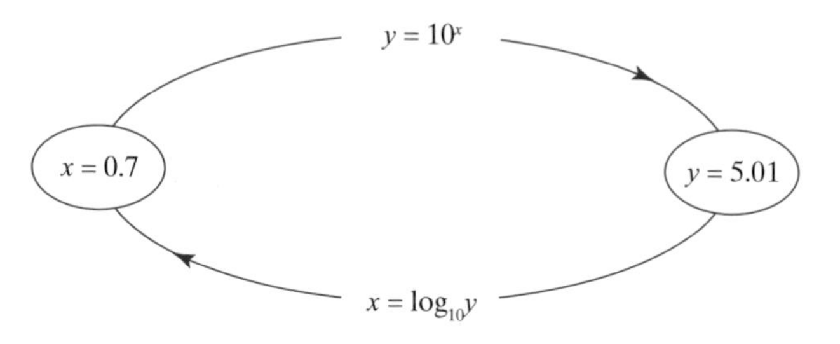

Common logarithms

Common logarithms are logarithms to base 10. They are one of the two systems of logarithms that can be found on all scientific calculators. We can use any letters and numbers when stating an exponential function. Previously we have used $a^x = y$. With common logarithms we let 'a' = 10 and 'y' = n so if $10^x = n$ then $x = \log_{10} n$.

- Find the keys on your calculator labelled as 10^x and LOG (meaning logarithm to the base 10).
- Evaluate $10^{4.2}$ by putting $x = 4.2$
- You should get 15 848.93192
- Now use the key labelled LOG and evaluate LOG 15 848.93192
- You should get back to 4.2

We can now solve the problem of finding x where $10^x = 4.6$ since $x = \log_{10} 4.6$

Use the log key on the calculator gives $x = 0.662758$ (to 6 d.p.).

Example
Find the values of the following logarithms without using a calculator:

a) $\log_{10} 10$
b) $\log_{10} 100$
c) $\log_{10} 1000$
d) $\log_{10} 1$

Answer

a) $x = \log_{10} 10$ so $10^x = 10$ so $x = 1$
b) $x = \log_{10} 100$ so $10^x = 100 = 10^2$ so $x = 2$
c) $x = \log_{10} 1000$ so $10^x = 1000 = 10^3$ so $x = 3$
d) $x = \log_{10} 1$ so $10^x = 1 = 10^0$ so $x = 0$

The key points from this example are:

$\log_{10} 10^n = n$

$\log_{10} 1 = 0$

$\log_{10} 10 = 1$

Essential notes

You must learn these key points.

Example
Use your calculator to find the values of the following logarithms to 4 significant figures

a) $\log_{10} 20$
b) $\log_{10} 0.5$
c) $\log_{10} 5.73$
d) $\log_{10} 0.12$

Answer

a) $\log_{10} 20 = 1.301$
b) $\log_{10} 0.5 = -0.3010$
c) $\log_{10} 5.73 = 0.7582$
d) $\log_{10} 0.12 = -0.9208$

Method notes

Use the **LOG** key on your calculator to answer the questions.

From these answers we see that if $n > 1$ then $\log_{10} n$ is positive and if $0 < n < 1$ then $\log_{10} n$ is negative.

Example
Solve the following equations

a) $10^x = 13$
b) $\log_{10} x = 2.1$

Answer

a) $10^x = 13$. The equivalent statement is $x = \log_{10} 13$ and using the log key gives $x = 1.1139$
b) $\log_{10} x = 2.1$. The equivalent statement is $x = 10^{2.1}$ and using the Λ key or the 10^x key gives $x = 125.89$

Logarithms to other bases

There are many other base numbers which we can use for logarithms as the following examples show.

Example

Write the following as logarithms.

a) $32 = 2^5$
b) $81 = 3^4$
c) $4^{-3} = \frac{1}{64}$

Answer

Write the equivalent statement.

a) $32 = 2^5$ so $\log_2 32 = 5$
b) $81 = 3^4$ so $\log_3 81 = 4$
c) $4^{-3} = \frac{1}{64}$ so $\log_4 \frac{1}{64} = -3$

Example

Write each of the following in index form.

a) $\log_2 128 = 7$
b) $\log_{10} 1\,000\,000 = 6$
c) $\log_3 243 = 5$
d) $\log_a 3 = b$

Answer

As in the example above always write the equivalent statement.

a) $\log_2 128 = 7$ so $2^7 = 128$
b) $\log_{10} 1\,000\,000 = 6$ so $10^6 = 1\,000\,000$
c) $\log_3 243 = 5$ so $3^5 = 243$
d) $\log_a 3 = b$ so $a^b = 3$ (the same rules apply in writing the equivalent statement whether we are given letters or numbers in the question.)

Method notes

When answering any questions involving exponentials and logarithms always write down the equivalent statement to the one you are given in the question.

If you are given an index statement write down the logarithmic statement.

If you are given a logarithmic statement write down the index statement.

It will then become clear which key you need to use on your calculator to answer the question.

Stop and think 1

What happens if you try to find the logarithm of a negative number e.g. $\log_{10}(-6)$?

Laws of logarithms

Two important results for logarithms are

$\log_a 1 = 0$

$\log_a a = 1$

where $a \neq 0$

Essential notes

If $a^m = 1$ then $m = 0$ so $\log_a 1 = 0$

If $a^m = a$ then $m = 1$ so $\log_a a = 1$

We have already seen the second of these results when $a = 10$ as $\log_{10} 10 = 1$

We now need to be able to use the standard mathematical operations of multiplication and division with the logarithmic function if we want to simplify statements such as $\log_3 12 - \log_3 16 + \log_3 7$. The following two laws will help us to do this.

The product law for logarithms

To work out the logarithm of two 'quantities' multiplied together we add the logarithm of the first 'quantity' to the logarithm of the second 'quantity'.

We can show this by letting $x = \log_a b$ where b is our first 'quantity' and letting $y = \log_a c$ where c is the second 'quantity'.

Proof

Step 1: If $x = \log_a b$ and $y = \log_a c$ the equivalent statements are $b = a^x$ and $c = a^y$

Step 2: Use the results from step 1 to give

$$b \times c = a^x \times a^y = a^{x+y} \text{ so } b \times c = a^{x+y}$$

Step 3: Use the equivalent statement for step 2 to give

$$\log_a b \times c = x + y$$

Step 4: Substitute $x = \log_a b$ and $y = \log_a c$ (from step1) into step 3 to give

$$\log_a b \times c = \log_a b + \log_a c$$

This shows that the logarithm of a **product** of numbers is the **sum** of the logarithms and the **product law** is:

$$\log_a b \times c = \log_a b + \log_a c$$

Essential notes

Remember the index law from Core 1: $a^x \times a^y = a^{x+y}$

The quotient law for logarithms

To work out the logarithm of one 'quantity' divided by another 'quantity' we subtract the logarithm of the 'dividing quantity' from the logarithm of the other 'quantity'.

We can show this by letting $x = \log_a b$ where b is one 'quantity' and letting $y = \log_a c$ where 'c' is the other 'quantity'.

Proof

Step 1: If $x = \log_a b$ and $y = \log_a c$ the equivalent statements are $b = a^x$ and $c = a^y$

Step 2: Use results from step 1 to give $\frac{b}{c} = \frac{a^x}{a^y} = a^{x-y}$ so $\frac{b}{c} = a^{x-y}$

Step 3: Use the equivalent statement for step 2 to give $x - y = \log_a \frac{b}{c}$

Step 4: Substitute $x = \log_a b$ and $y = \log_a c$ (from step 1) into step 3 to give

$$\log_a b - \log_a c = \log_a \frac{b}{c}$$

This shows that the logarithm of a **quotient** of numbers is the **difference** of the logarithms and the **quotient law** is:

$$\log_a \frac{b}{c} = \log_a b - \log_a c$$

From this law we can deduce the following very useful result:

$$\log_a \frac{1}{c} = \log_a 1 - \log_a c = -\log_a c$$

Essential notes

From Core 1: $\frac{1}{c} = c^{-1}$

Remember that $\log_a 1 = 0$

If we are asked to solve the equation $4^{x+5} = 3^{4-3x}$ we need to know how to use powers or indices with the logarithmic function. The following law will enable us to do that.

The power law for logarithms

If you have a 'quantity raised to a power' and you then take the logarithm of that 'quantity raised to a power' this is the same as multiplying the power by the logarithm of the original quantity before it was raised to a power. We can show this by letting $x = \log_a b$ where b is the 'original quantity'.

Essential notes

Remember the index law from Core 1: $(a^x)^n = a^{nx}$.

Proof

Step 1: If we let $x = \log_a b$ the equivalent statement is $b = a^x$.

Step 2: Use algebra $b^n = (a^x)^n = a^{nx}$ so $b^n = a^{nx}$

Step 3: The equivalent statement is $\log_a b^n = nx$

Step 4: Substitute for x from step 1 to give $\log_a b^n = n \log_a b$

This is the power law for logarithms.

$\log_a b^n = n \log_a b$

Exam tips

You must **learn** these three laws and their proofs. They will not be given in the formula book.

Product: $\log_a (b \times c) = \log_a b + \log_a c$

Quotient: $\log_a \left(\frac{b}{c}\right) = \log_a b - \log_a c$

Power: $\log_a b^n = n \log_a b$

Example

Write each of the following as a single logarithm.

a) $\log_3 12 - \log_3 16 + \log_3 7$
b) $2\log_{10} x + \log_{10} xy - 3\log_{10} y$

Answer

a) $\log_3 12 - \log_3 16 + \log_3 7 = \log_3 \frac{12 \times 7}{16} = \log_3 \frac{21}{4}$
b) $2\log_{10} x + \log_{10} xy - 3\log_{10} y = \log_{10} x^2 + \log_{10} xy - \log_{10} y^3$
$= \log_{10} \frac{x^2 \times xy}{y^3} = \log_{10} \frac{x^3}{y^2}$

Method notes

a) Use the product law for $\log_3 12 + \log_3 7$ to give $\log_3 12 \times 7$ then the quotient law for $\log_3 12 \times 7 - \log_3 16 = \log_3 \frac{12 \times 7}{16}$
Simplifying this numerically $= \log_3 \frac{21}{4}$
b) Use the power law to rewrite $2\log_{10} x$ and $3\log_{10} y$ as $\log_{10} x^2$ and $\log_{10} y^3$.

Then proceed as in (a) using the product and quotient laws.

Example

Expand the following in terms of $\log u$, $\log v$ and $\log w$

a) $\log_5 uv$
b) $\log_3 u^3 v^4$
c) $\log \frac{u^2\sqrt{v}}{w^5}$

Answer

a) $\log_5 uv = \log_5 u + \log_5 v$ (using product law)
b) $\log_3 u^3 v^4 = \log_3 u^3 + \log_3 v^4 = 3\log_3 u + 4\log_3 v$ (using product and power laws)

c) $\log \frac{u^2\sqrt{v}}{w^5} = \log u^2 + \log\sqrt{v} - \log w^5$

$= 2\log u + \frac{1}{2}\log v - 5\log w$

(using product, quotient and power laws)

Method notes

Remember $\sqrt{v} = v^{\frac{1}{2}}$

Solution of equations of the form $a^x = b$

Earlier in this chapter we have seen how to solve the equation $y = 10^x$ when $y = 4.6$ using common logarithms. An equation of this type is called an **exponential equation** of the general form $a^x = b$. In this case the base number is 10. The solution was $x = \log_{10}4.6 = 0.662758$ (to 6 d.p.)

There are occasions when solutions are required to exponential equations involving bases other than 10. We can often solve such equations by 'taking logs to base 10' throughout the equation and then using the rules of logarithms as explained in the examples below.

Example

Solve the following equations giving the answers to 3 significant figures.

a) $2^x = 27$
b) $4^{x+5} = 3^{4-3x}$
c) $3^{2x} - 6(3^x) + 5 = 0$

Answer

a) Solve $2^x = 27$

Step 1: Take logs to base 10 of both sides of the equation to give
$\log_{10} 2^x = \log_{10} 27$

Step 2: Use the power law to give $x \log_{10} 2 = \log_{10} 27$

Step 3: Divide both sides by $\log_{10} 2$ (this is a constant) to give
$x = \frac{\log_{10} 27}{\log_{10} 2}$ so $x = \frac{1.43136}{0.30103} = 4.75$ (to 3 s.f.) from calculator

b) Solve $4^{x+5} = 3^{4-3x}$

Step 1: Take logs to base 10 of both sides of the equation to give
$\log_{10} 4^{x+5} = \log_{10} 3^{4-3x}$

Step 2: Use the power law to give $(x+5)\log_{10} 4 = (4-3x)\log_{10} 3$

Step 3: Multiply to give $x \log_{10} 4 + 5\log_{10} 4 = 4\log_{10} 3 - 3x\log_{10} 3$

Step 4: Simplify to give $x\log_{10} 4 + 3x\log_{10} 3 = 4\log_{10} 3 - 5\log_{10} 4$

Step 5: Factorise to give $x(\log_{10} 4 + 3\log_{10} 3) = 4\log_{10} 3 - 5\log_{10} 4$

Step 6: Simplify to give $x = \frac{4\log_{10}3 - 5\log_{10}4}{\log_{10}4 + 3\log_{10}3} = -0.542$ from calculator

Method notes

Logarithms to base 10 are available on all scientific calculators. The method is called 'taking logs to base 10 of each side'.

Method notes

You will need to learn the method for solving this type of equation where different powers of x appear. By substituting correctly (in this case let $y = 3^x$) you obtain a quadratic equation which can then be solved in the usual way.

c) Solve $3^{2x} - 6(3^x) + 5 = 0$

Step 1: From index rules $3^{2x} = (3^x)^2$ and let $y = 3^x$

Step 2: Substitute for y from step 1 to give $3^{2x} = y^2$

Step 3: Substitute for y and y^2 from step 2 to give $3^{2x} - 6(3^x) + 5$
$= y^2 - 6y + 5 = 0$

Step 4: Solve the quadratic to give $(y - 5)(y - 1) = 0$ so $y = 5$ or $y = 1$

Step 5: If $y = 5$ then from step 1: $3^x = 5$

Step 6: Use the 'logs to base 10' method to give $\log_{10} 3^x = \log_{10} 5$

Step 7: Use the power law to give $x\log_{10} 3 = \log_{10} 5$

Step 8: Simplify to give $x = \dfrac{\log_{10} 5}{\log_{10} 3} = 1.46$ (to 3 s.f.)

Step 9: From step 4: if $y = 1$ then $3^x = 1$ so $x = 0$ (index rule)

The change of base law for logarithms

We often need to change the base of the logarithm to another base. We can use the change of base law which is

$$\log_a n = \frac{\log_b n}{\log_b a}$$

Method notes

Step 2: you have seen this method before with $b = 10$

Exam tips

You must **learn** the change of base law and its proof. It will not be given in the formula book.

$$\log_a n = \frac{\log_b n}{\log_b a}$$

Proof

Suppose that we wish to change $\log_a n$ to a logarithm to base b.

Step 1: Let $y = \log_a n$ then the equivalent statement is $n = a^y$.

Step 2: Take logarithms to base b of each side of the equivalent statement obtained in step 1 to give $\log_b n = \log_b a^y$.

Step 3: Use the power law of logarithms to give $\log_b n = y \log_b a$

Step 4: Simplify to give $y = \dfrac{\log_b n}{\log_b a}$

Step 5: Substitute for y from step 1 in step 4 to give $\log_a n = \dfrac{\log_b n}{\log_b a}$

Example

a) Find the value of $\log_5 7$ correct to 3 significant figures.

b) Solve the equation $\log_3 x + 6\log_x 3 = 5$ giving the answer to 3 significant figures.

Answer

a) **Step 1:** We do not have a calculator key for base 5. To answer the question we must rewrite it in terms of base 10

Step 2: Use the change of base law $\log_a n = \frac{\log_b n}{\log_b a}$ with $a = 5$, $n = 7$ and $b = 10$ to give $\log_5 7 = \frac{\log_{10} 7}{\log_{10} 5} = \frac{0.845098}{0.698970} = 1.21$

b) **Step 1:** We need to express both logarithms to base 3. Use the change of base rule with $n = 3$ and $a = x$ to give

$$\log_x 3 = \frac{\log_3 3}{\log_3 x} = \frac{1}{\log_3 x}$$

Step 2: Rewrite the question using step 1 to give $\log_3 x + \frac{6}{\log_3 x} = 5$

Step 3: If $y = \log_3 x$ in step 2 then $y + \frac{6}{y} = 5$

Step 4: Simplify: $y^2 - 5y + 6 = 0$ to give $(y - 3)(y - 2) = 0$

Step 5: Solve the quadratic for $y = 3$ or $y = 2$

Step 6: If $y = 3$ then $\log_3 x = 3$ so $x = 3^3 = 27$

Step 7: If $y = 2$ then $\log_3 x = 2$ so $x = 3^2 = 9$

Method notes

Another method is to write the equivalent statement $5^x = 7$ then take logs base 10 to give $\log_{10} 5^x = \log_{10} 7$

so $x\log_{10} 5 = \log_{10} 7$

$$x = \frac{\log_{10} 7}{\log_{10} 5} = 1.21$$

b) Different bases so we must rewrite the question so that we have the same base throughout.

Step 1: Remember that $\log_a a = 1$

Step 6: If $\log_3 x = 3$ the equivalent statement is $x = 3^3$

Stop and think answers

If you try to find the value of $\log_{10}(-6)$ on your calculator it will show 'math error'.

More generally let $y = \log_a(-n)$ then the equivalent statement is $-n = a^y$.

As a is the base of the logarithm it is positive.

If y is positive then a^y is positive.

If y is negative then by rules of indices $a^{\text{negative } y}$ means $\frac{1}{a^{\text{positive } y}}$ which will always be positive.

We can therefore conclude that there is no (real) number y and base a for which a^y is negative so $\log_a(-n)$ does not give a real number for any base a.

Essential notes

This process is called 'differentiating with respect to x' or 'finding the derivative with respect to x'.

Differentiation was introduced in Core 1. The following points summarise what you are expected to know about this process in order to move on to differentiation in Core 2.

The rule of differentiation

Given $y = f(x) = x^n$ then $\frac{dy}{dx} = f'(x) = nx^{n-1}$

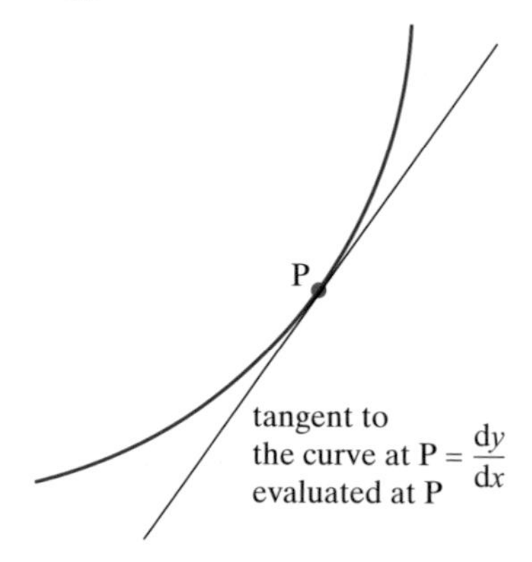

Fig. 6.1
Gradient of the tangent to the curve at P = $\frac{dy}{dx}$ evaluated at P.

- $\frac{dy}{dx}$ or $f'(x)$ is called the gradient function and describes the rate of change of y with respect to x.
- The value of $\frac{dy}{dx}$ or $f'(x)$ at a point on the curve $y = f(x)$ measures the slope or **gradient** of the tangent to the graph at that point.
- $\frac{d^2y}{dx^2}$ or $f''(x)$ is called the second derivative of y with respect to x.
- $\frac{d^2y}{dx^2}$ is found by differentiating $\frac{dy}{dx}$ with respect to x.
- $f''(x)$ is found by differentiating $f'(x)$ with respect to x.

Example

a) Find the first derivative $\frac{dy}{dx}$ and the second derivative $\frac{d^2y}{dx^2}$ of the function $y = x^4 - 3x^2 + 5x - 2$

b) Find the gradient of the tangent to the curve with equation $y = x^4 - 3x^2 + 5x - 2$ at the point (2, 12).

Answer

a) $y = x^4 - 3x^2 + 5x - 2$ differentiating gives $\frac{dy}{dx} = 4x^3 - 6x + 5$

differentiating $\frac{dy}{dx}$ gives $\frac{d^2y}{dx^2} = 12x^2 - 6$

b) $\frac{dy}{dx}$ gives the gradient of the tangent at any point.

At the point (2, 12) $\frac{dy}{dx} = 4 \times 2^3 - 6 \times 2 + 5 = 25$ so the gradient of the tangent at (2, 12) is 25

Increasing and decreasing functions

In Core 1 we used the equation of a straight line

$y = mx + c$ where the value of m describes the gradient of the line and c is a **constant**.

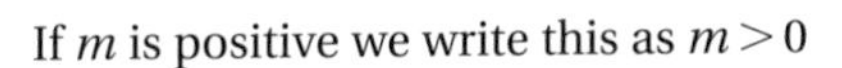

If m is positive we write this as $m > 0$

For example, if $m = 3$ in $y = mx + c$:

if $x = 1$ then $y = 1(3) + c$: $y = 3 + c$

if $x = 2$ then $y = 2(3) + c$: $y = 6 + c$

This shows that y increases as x increases when $m > 0$

This result is shown graphically in Fig 6.2a.

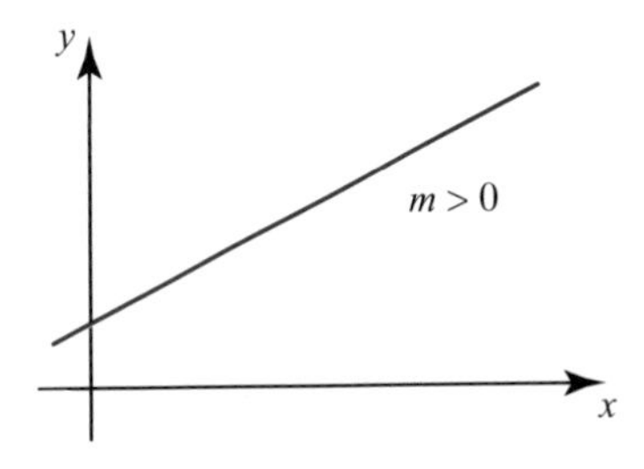

Fig. 6.2 a

If m is negative we write this as $m < 0$

For example, if $m = -2$ in $y = mx + c$:

if $x = 1$ then $y = 1(-2) + c$: $y = -2 + c$

if $x = 2$ then $y = 2(-2) + c$: $y = -4 + c$

This shows that y decreases as x increases when $m < 0$

This result is shown graphically in Fig 6.2b.

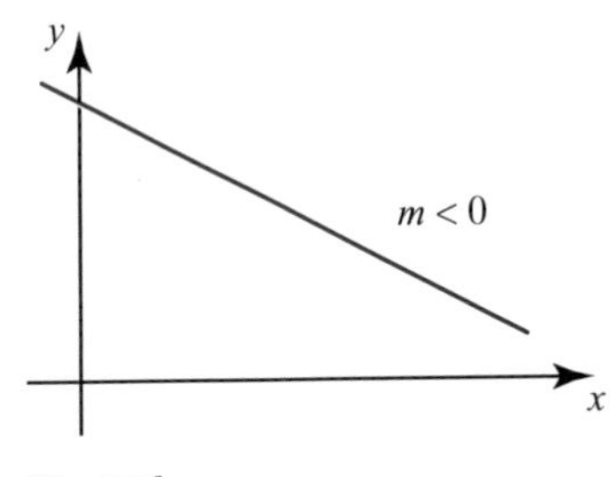

Fig. 6.2 b

General definition of an increasing function

A function $f(x)$ is an **increasing** function if, when x is a given set of values taken from $\{x_0, x_1, x_2, \dots, x_n\}$ and $x_1 < x_2$, the outcome is $f(x_1) < f(x_2)$. This is shown graphically in Fig 6.3 below.

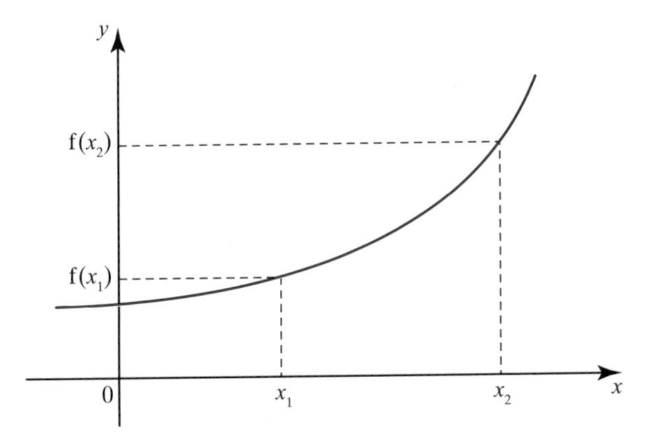

Fig. 6.3
An increasing function.

Essential notes

Figure 6.3 is an example of an increasing function. The slope of the tangent at each point is positive.

We can also see from the graph that a function f is increasing if at all points on its graph the gradient of the tangent is positive, that is $f'(x) > 0$ (remember that $f'(x)$ is the gradient function).

Method notes

The test for an increasing function is that $f'(x) > 0$

Boundary values were covered in Core 1 in solving quadratic inequalities.

Example

Find the values of x for which $f(x) = x^3 - 9x^2 - 21x$ is an increasing function.

Answer

Step 1: Differentiate $f(x) = x^3 - 9x^2 - 21x$ to give $f'(x) = 3x^2 - 18x - 21$

Step 2: Solve $3x^2 - 18x - 21 > 0$

Step 3: Divide by 3 to give $x^2 - 6x - 7 > 0$

Step 4: Factorise to give $(x-7)(x+1) > 0$ giving 7 and -1 as the boundary values.

Step 5: $f'(x) > 0$ when $x < -1$ and $x > 7$

Therefore the function $f(x) = x^3 - 9x^2 - 21x$ is increasing for $x < -1$ and $x > 7$

Definition of a decreasing function

A function $f(x)$ is an **decreasing** function if, when x is a given set of values taken from $\{x_0, x_1, x_2, \ldots, x_n\}$ and $x_1 < x_2$, the outcome is $f(x_1) > f(x_2)$. This is shown graphically in Fig 6.4 below.

We can also see from the graph that a function f is decreasing if at all points on its graph the slope of the tangent is negative, that is $f'(x) < 0$

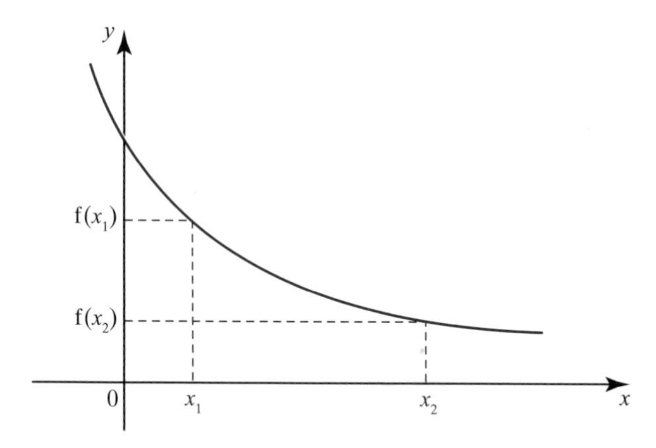

Fig. 6.4
A decreasing function.

Essential notes

Figure 6.4 is an example of a decreasing function. The slope of the tangent at each point is negative.

Method notes

The test for a decreasing function is that $f'(x) < 0$.

Boundary values were covered in Core 1 in solving quadratic inequalities.

Example

Find the values of x for which $f(x) = x^4 - 8x^3$ is a decreasing function.

Answer

Step 1: Differentiate $f(x) = x^4 - 8x^3$ to give $f'(x) = 4x^3 - 24x^2$

Step 2: Solve $4x^3 - 24x^2 < 0$

Step 3: Divide by 4 to give $x^3 - 6x^2 < 0$

Step 4: Factorise $x^2(x-6) < 0$ giving 0 and 6 as the boundary values

Step 5: $f'(x) < 0$ when $x < 0$ and $x < 6$

Therefore the function $f(x) = x^4 - 8x^3$ is decreasing for $x < 0$ and $x < 6$

This is the same as the single statement $x < 6$ because if $x < 0$ x is also < 6

Stationary points of a graph

Example

Describe how the gradient changes as you move along the curve at the points A, B, C, D and E of the graph in Figure 6.5 and what this tells us about the function.

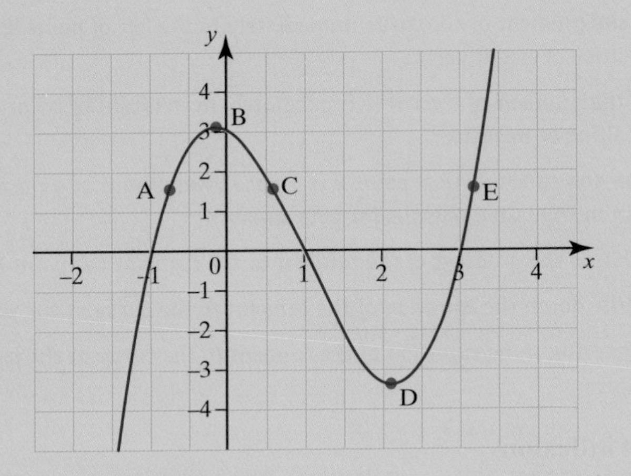

Fig. 6.5

Answer

The gradient of the graph changes as you move along it.

- at A the gradient is positive therefore the function is increasing
- at B the gradient is zero; this point is called a **(local) maximum**
- at C the gradient is negative therefore the function is decreasing
- at D the gradient is zero; this point is called a **(local) minimum**
- at E the gradient is positive therefore the function is increasing.

Essential notes

Often you will see these points described as just maximum or minimum points.

Strictly speaking this is not true because there are points on the curve which are higher than the B (some to the right of point E) and points that are lower than D (some to the left of point A).

In the above example the points B and D are examples of **stationary points** or **turning points** because

- At B the graph is increasing for points to the left of B, decreasing for the points to the right of B.
- At D the graph is decreasing for points to the left of D, increasing for the points to the right of D.

The tangent at each point B and D is horizontal.

Definition

Points on the curve of a function $y = \text{f}(x)$ where $\frac{\text{d}y}{\text{d}x} = \text{f}'(x) = 0$ are called stationary points or turning points.

Maxima and minima of functions

- At a (local) **maximum** point the function changes from increasing (positive gradient to the left of the point) to decreasing (negative gradient to the right of the point).

- At a (local) **minimum** point the function changes from decreasing (negative gradient to the left of the point) to increasing (positive gradient to the right of the point).

Stop and think 1

These questions refer to the graph in Figure 6.5.

1. a) *Is the gradient of the curve immediately to the left of point B, positive or negative?*
   b) *Is the gradient of the curve immediately to the right of point B positive or negative?*
   c) *Can you conclude that point B is a (local) maximum or a (local minimum)? Give reasons for your answer.*
2. a) *What is the gradient of the tangent to the curve at the point B?*
   b) *Write down the equation of the tangent to the curve at the point B.*
   c) *Write down the equation of the tangent to the curve at the point D.*

Essential notes

There are points of inflexion where the gradient is not zero but this is not explored in Core 2.

Points of inflexion

Definition

If at a stationary point the function is either increasing on each side of that stationary point or decreasing on each side of that stationary point then the point is called a **point of inflexion**. This is shown in Figure 6.6 below.

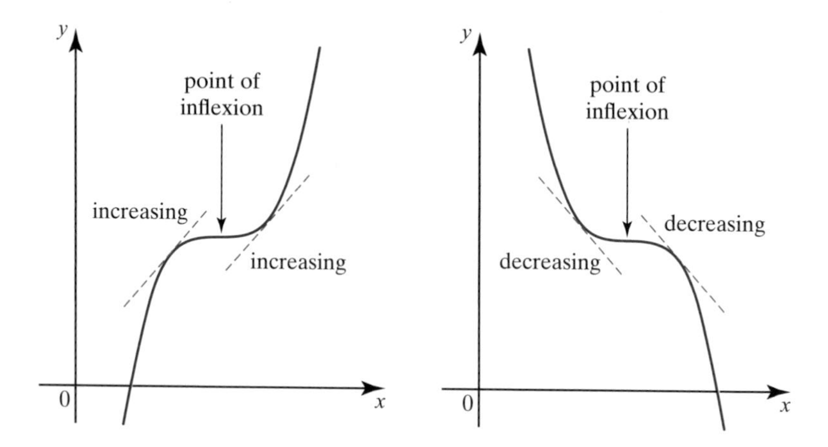

Fig. 6.6
Points of inflexion.

Example

Find the coordinates of the stationary points for the curve $y = x^3(4 - x)$. Classify each stationary point.

Answer

Step 1: Differentiate $y = 4x^3 - x^4$ so $\frac{dy}{dx} = 12x^2 - 4x^3$

Step 2: At a stationary point $\frac{dy}{dx} = 0$ so $12x^2 - 4x^3 = 0$

Step 3: Factorise to give $4x^2(3-x)=0$ so $x=0$ or $x=3$

Step 4: Work out the gradient either side of the stationary points:

value of x	$x = -0.1$	$x = 0$	$x = 0.1$
gradient	$\frac{dy}{dx} = 0.124$	$\frac{dy}{dx} = 0$	$\frac{dy}{dx} = 0.116$
slope of tangent	/	—	/
value of x	$x = 2.9$	$x = 3$	$x = 3.1$
gradient	$\frac{dy}{dx} = 3.364$	$\frac{dy}{dx} = 0$	$\frac{dy}{dx} = -3.844$
slope of tangent	/	—	\

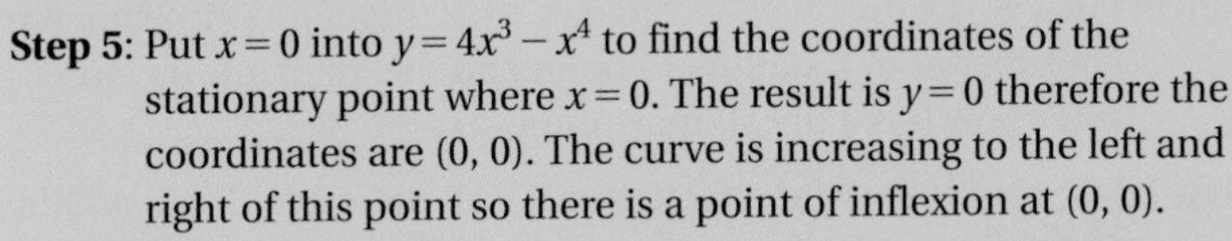

Step 5: Put $x=0$ into $y=4x^3-x^4$ to find the coordinates of the stationary point where $x=0$. The result is $y=0$ therefore the coordinates are (0, 0). The curve is increasing to the left and right of this point so there is a point of inflexion at (0, 0).

Step 6: Put $x=3$ into $y=4x^3-x^4$ to find the coordinates of the stationary point where $x=3$. The result is $y=27$ therefore the coordinates are (3, 27). The curve is increasing to the left and decreasing to right of this point so there is a (local) maximum at (3, 27).

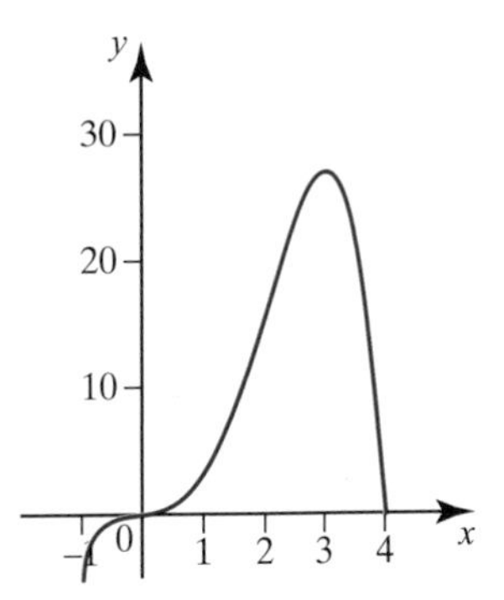

Fig. 6.7
The graph of $y=x^3(4-x)$.

Classifying stationary points using second derivatives

The second derivative $\frac{d^2y}{dx^2}$ can also be used to find out if a stationary point is a (local) maximum or (local) minimum.

Example

a) Find the coordinates of the stationary points of the function $y=x^3-3x$. Classify each stationary point.

b) Find the value of $\frac{d^2y}{dx^2}$ at each stationary point.

Answer

a) **Step 1**: Differentiate $y=x^3-3x$ so $\frac{dy}{dx}=3x^2-3$

Step 2: At a stationary point $\frac{dy}{dx}=0$ so $\frac{dy}{dx}=3x^2-3=0$

Step 3: Factorise to give $3(x^2-1)=3(x+1)(x-1)=0$

Method notes

x^2-1 is factorised using the difference of two squares formula

$a^2-b^2=(a+b)(a-b)$ where $a=x$ and $b=1$

Step 4: Solve for x to give $x = -1$ or $x = 1$

Step 5: Work out the gradient either side of the stationary points:

value of x	$x = -1.1$	$x = -1$	$x = -0.9$
gradient	$\frac{dy}{dx} = 0.63$	$\frac{dy}{dx} = 0$	$\frac{dy}{dx} = -0.57$
slope of tangent	/	—	\
value of x	$x = 0.9$	$x = 1$	$x = 1.1$
gradient	$\frac{dy}{dx} = -0.57$	$\frac{dy}{dx} = 0$	$\frac{dy}{dx} = 0.63$
slope of tangent	\	—	/

Method notes

Function is increasing to the left of (−1, 2) and decreasing to the right of the point.

Function is decreasing to the left of (1, −2) and increasing to the right of the point.

Step 6: Put $x = -1$ into $y = x^3 - 3x$ to give $y = 2$ therefore (−1, 2) are the coordinates of a stationary point and the gradients on either side show it is a (local) maximum.

Step 7: Put $x = 1$ into $y = x^3 - 3x$ to give $y = -2$ therefore (1, −2) are the coordinates of a stationary point and the gradients on either side show it is a (local) minimum.

b) **Step 1**: Differentiate $\frac{dy}{dx} = 3x^2 - 3$ to give $\frac{d^2y}{dx^2} = 6x$

Step 2: At the ((local) **maximum** point (−1, 2), $x = -1$ so $\frac{d^2y}{dx^2} = 6x = -6$ which is **negative**.

Step 3: At the (local) **minimum** point (1, −2), $x = +1$ so $\frac{d^2y}{dx^2} = 6x = +6$ which is **positive**

Part b) above demonstrates a useful method for classifying stationary points using second derivatives.

Essential notes

You must learn how to use the second derivative test for your examination.

General statement of the second derivative test

If $(a, f(a))$ are the coordinates of the stationary point of a function $y = f(x)$ then:

- $\frac{dy}{dx} = f'(a) = 0$ and $\frac{d^2y}{dx^2} = f''(a) < 0$ at a (local) maximum
- $\frac{dy}{dx} = f'(a) = 0$ and $\frac{d^2y}{dx^2} = f''(a) > 0$ at a (local) minimum

Special case of the second derivative test

Caution is needed if the first and second derivatives are both zero at some point on the curve $(a, f(a))$.

If $\frac{dy}{dx} = f'(a) = 0$ and $\frac{d^2y}{dx^2} = f''(a) = 0$ then the point could be a maximum, a minimum or a point of inflexion.

In this special case in order to classify the stationary point you must use the earlier method of finding the gradient on either side of the stationary point.

Example

For the following functions use the second derivative test to determine the nature of any stationary points.

a) $f(x) = 3x^4 - 4x^3 - 36x^2$

b) $f(x) = 2x^2(3 - \sqrt{x})$

Answer

a) **Step 1:** Differentiate $f(x) = 3x^4 - 4x^3 - 36x^2$ so $f'(x) = 12x^3 - 12x^2 - 72x$

Step 2: At stationary point $f'(x) = 0$ so $12x^3 - 12x^2 - 72x = 0$

Step 3: Solve $12x(x^2 - x - 6) = 0$ so $12x(x - 3)(x + 2) = 0$ so $x = 0$, $x = 3$ or $x = -2$

Step 4: Differentiate $f'(x)$ so $f''(x) = 36x^2 - 24x - 72$

Step 5: Find the value of $f''(x)$ when $x = -2$ to give $f''(x) = 36(-2)^2 - 24(-2) - 72 = 120 > 0$ which is positive so there is a local minimum when $x = -2$

Step 6: Repeat step 5 when $x = 0$ to give $f''(x) = 36(0)^2 - 24(0) - 72 = -72 < 0$ which is negative so there is a local maximum when $x = 0$

Step 7: Repeat step 5 when $x = 3$ to give $f''(x) = 36(3)^2 - 24(3) - 72 = 180 > 0$ which is positive so there is a local minimum when $x = 3$

b) **Step 1:** Simplify $f(x) = 2x^2(3 - \sqrt{x}) = 6x^2 - 2x^{2.5}$

Step 2: Differentiate $f'(x) = 12x - 5x^{1.5} = x(12 - 5\sqrt{x})$

Step 3: At a stationary point $f'(x) = 0$ so $x(12 - 5\sqrt{x}) = 0$

Step 4: Solve $x(12 - 5\sqrt{x}) = 0$ to give $x = 0$ or $12 - 5\sqrt{x} = 0$ so $x = \frac{144}{25}$

Step 5: Differentiate $f'(x)$ so $f''(x) = 12 - 7.5x^{0.5} = 12 - 7.5\sqrt{x}$

Step 6: Find the value of $f''(x)$ when $x = 0$ to give $f''(0) = 12 - 7.5\sqrt{0} = 12 > 0$ which is positive so the stationary point is a local minimum.

Step 7: Repeat step 6 when $x = \frac{144}{25}$ to give $f''(x) = 12 - 7.5\sqrt{\frac{144}{25}} = -6 < 0$ which is negative so the stationary point is a local maximum.

Method notes

If you are asked for the coordinates of any stationary points you must work out the y value at each of the points. In this example if you had been asked for them in (a) $x = 0$ so

$y = f(x) = 3(0) - 4(0)^3 - 36(0)^2 = 0$ so there is a maximum point at $(0, 0)$

Similarly there is a minimum point at $(-2, -64)$ and a minimum point at $(3, -189)$.

Method notes

Simplify $f(x)$ using the rules of indices (covered in Core 1).

If $12 - 5\sqrt{x} = 0$ this means $12 = 5\sqrt{x}$ and squaring both sides of this equation gives $144 = 25x$ so $\frac{144}{25} = x$.

Example

Show that the point (2, 3) is a point of inflexion of the function $f(x) = x^3 - 6x^2 + 12x - 5$

Answer

Step 1: Differentiate $f(x) = x^3 - 6x^2 + 12x - 5$ to give $f'(x) = 3x^2 - 12x + 12$

Step 2: At a stationary point $f'(x) = 0$ so $3x^2 - 12x + 12 = 0$

Step 3: Solve so $3(x^2 - 4x + 4) = 0$ which means $3(x - 2)^2 = 0$ so $x = 2$

Step 4: Differentiate $f(x)$ to give $f''(x) = 6x - 12$

Step 5: Find the value of $f''(x)$ when $x = 2$, so $f''(2) = 6(2) - 12 = 0$ so the stationary point cannot be classified using the second derivative test.

Step 6: To classify the stationary point use the method of looking at the gradient either side of the point.

value of x	$x = 1.9$	$x = 2$	$x = 2.1$
gradient	$\frac{dy}{dx} = 0.03 > 0$	$\frac{dy}{dx} = 0$	$\frac{dy}{dx} = 0.03 > 0$
slope of tangent	/	—	\

The curve is increasing to the left of the point where $x = 2$ and increasing to the right of the point where $x = 2$ so there is a point of inflexion at the point (2, 3) as shown in Figure 6.8.

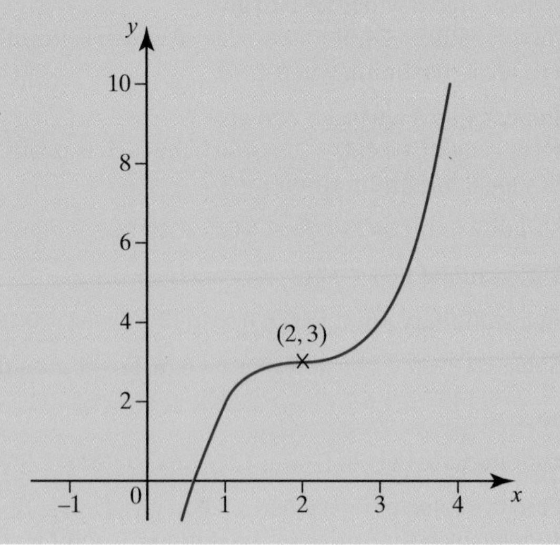

Fig. 6.8
A sketch of the function $f(x) = x^3 - 6x^2 + 12x - 5$ showing the point of inflexion at (2, 3).

Method notes

You should check that the point given is on the curve. If $x = 2$, $f(x) = y = (2)^3 - 6(2)^2 + 12(2) - 5 = 3$ so (2, 3) is a point on this curve.

Application of maxima and minima in problem solving

Optimisation

In many real life problems the aim is to find the maximum or minimum of a function. For example a company may wish to maximise its profits or a manufacturer may wish to minimise the amount of material used in making an object. These are called **optimisation** problems.

Example

A window has a semi-circular arch (of radius x metres) above a rectangular section as shown in Figure 6.9. The perimeter of the window must be 6 m.

a) Show that $y = 3 - \frac{1}{2}\pi x$

b) Show that the area of the window is **maximised** when $x = \frac{6}{\pi + 4}$

c) Find the **maximum area** of the window.

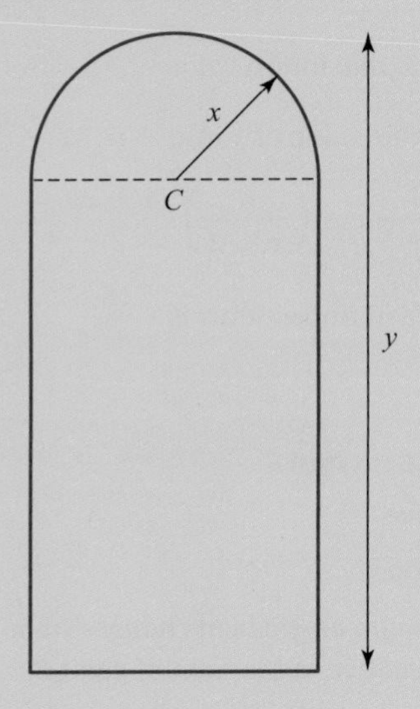

Fig. 6.9
Diagram of the window

Answer

a) **Step 1**: Let the perimeter be represented by P metres and work out the algebraic formula for the perimeter.

P = length of the 3 sides of the rectangle + perimeter of the semi-circular arch.

$= 2(y - x) + 2x + \pi x$ so $P = 2y + \pi x$

Step 2: We are told that $P = 6$ so $2y + \pi x = 6$ so $y = 3 - \frac{1}{2}\pi x$ (as required) (1)

b) **Step 1**: Let the area of the window be represented by A square metres and work out the algebraic formula for A.

Essential notes

Use the second derivative test to classify the stationary point. Here $\frac{d^2A}{dx^2} = -(4+\pi)$ will always be negative whatever the value of x since 4 and π are both constant and x does not appear in $\frac{d^2A}{dx^2} = -(4+\pi)$.

Step 2: A = area of semi-circle + area of rectangle

$$A = \frac{1}{2}\pi x^2 + 2x(y - x) \qquad (2)$$

Step 3: Substitute for y from equation (1) into equation (2) to give

$$A = \frac{1}{2}\pi x^2 + 2x\left(3 - \frac{1}{2}\pi x - x\right) = 6x - 2x^2 - \frac{1}{2}\pi x^2$$

Step 4: Simplify the algebra to give $A = 6x - \frac{x^2}{2}(4 + \pi)$

Step 5: A maximum value of the area so $\frac{dA}{dx} = 0$ (3)

Step 6: Differentiate A so $\frac{dA}{dx} = 6 - x(4 + \pi)$ (4)

Step 7: Combine equation (3) and (4) to give $0 = 6 - x(4 + \pi)$ so

$$x = \frac{6}{\pi + 4}$$

Step 8: Differentiate to give $\frac{d^2A}{dx^2} = -(4+\pi)$. This is **always** negative

so if $x = \frac{6}{\pi + 4}$ the stationary point is maximum.

c) **Step 1:** From (b) the maximum value of A occurs when $x = \frac{6}{\pi + 4}$

Step 2: Substitute this value of x into $A = 6x - \frac{x^2}{2}(4 + \pi)$ to give

$$A = 6\left(\frac{6}{\pi + 4}\right) - \left(\frac{6}{\pi + 4}\right)^2 \frac{(4 + \pi)}{2} = \frac{18}{\pi + 4}$$

The maximum area of the window is $\frac{18}{\pi + 4}$

Stop and think answers

1 a) *Gradient is positive*

b) *Gradient is negative*

c) *It is a local maximum as gradient changes from positive to the left of point B then negative to the right of point B.*

2 a) *Gradient of the tangent at point B is 0 because the tangent is horizontal at point B.*

b) *Using $y = mx + c$ as the equation of the straight line (which is the tangent at the point B) gradient $m = 0$ so $y = 0x + c$ so $y = c$ is the required equation.*

(From the graph an approximate value for c could be 3 so from the information given $y = 3$ is also an acceptable answer.)

c) *Similarly using the same method as in b) $y = d$ (where d is a constant) or $y = -3.2$ are acceptable answers.*

Definite integration

In Core 1 integration was used to find the area under a graph by considering the area under the graph of $y = mx + c$ between $x = a$ and $x = b$ as illustrated in Figure 7.1 below.

Fig. 7.1
The area under the graph of $y = mx + c$ between $x = a$ and $x = b$.

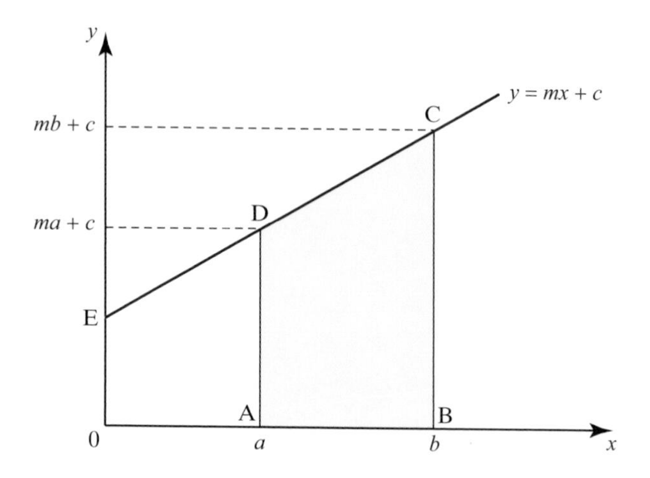

Essential notes

OA = a OB = b

y coordinate of the point D = $ma + c$ since D is on the line $y = mx + c$.

The area of the trapezium OADE is $\frac{1}{2}(c + (ma + c))a$.

The area of the trapezium OBCE is $\frac{1}{2}(c + (mb + c))b$.

Essential notes

E is the point where the line $y = mx + c$ intercepts the y-axis so length of OE is c.

Area of a trapezium = $\frac{1}{2}$(sum of the lengths of parallel sides) × perpendicular distance between them.

Subtracting the two expressions above shows that the area of the trapezium ABCD is:

$$\frac{1}{2}(c + (mb + c))b - \frac{1}{2}(c + (ma + c))a = \frac{1}{2}(2cb + mb^2) - \frac{1}{2}(2ca + ma^2)$$

$$= \left(\frac{1}{2}mb^2 + cb\right) - \left(\frac{1}{2}ma^2 + ca\right)$$

We can see that the expression in each bracket is of the same form. If we write $A(x) = \frac{1}{2}mx^2 + cx$, then $A(b) = \frac{1}{2}mb^2 + cb$ and $A(a) = \frac{1}{2}ma^2 + ca$ so area of ABCD = $A(b) - A(a)$.

The function $A(x)$ is called the area function and for the function $f(x) = mx + c$ (that is $y = mx + c$):

$$A(x) = \frac{1}{2}mx^2 + cx$$

Definition

As stated in Core 1 the area function $A(x)$ is also called the integral function. Hence $A(x)$ is the integral of the function $f(x)$.

This means that the finite area ABCD is written as:

$$\int_a^b f(x)dx = [A(x)]_a^b = A(b) - A(a)$$

and is called a **definite integral** since it has a finite value.

a is called the lower limit of the integral and b is called the upper limit of the integral. The use of square brackets indicates that there are upper and lower limits which must be used to find the finite answer.

Example

Evaluate the following definite integrals.

a) $\int_0^2 (x^3 + 2x^2 - 4x + 5)dx$

b) $\int_1^8 (x^{\frac{1}{3}} - 1)dx$

c) $\int_{-1}^2 (x - 2)^2 dx$

Answer

a) **Step 1**: Integrate and transfer the limits to the square brackets

$$\int_0^2 (x^3 + 2x^2 - 4x + 5)dx = \left[\frac{x^4}{4} + 2\frac{x^3}{3} - 2x^2 + 5x\right]_0^2$$

Step 2: Evaluate using the upper limit $x = 2$ and subtract the evaluation for $x = 0$: $\left[\frac{2^4}{4} + \frac{2^4}{3} - 2^3 + 10\right] - [0] = 11\frac{1}{3}$

Use the Method notes to answer (b) and (c).

b) $$\int_1^8 (x^{\frac{1}{3}} - 1)dx = \left[\frac{x^{\frac{4}{3}}}{\frac{4}{3}} - x\right]_1^8 = \left[\frac{3}{4} \times 8^{\frac{4}{3}} - 8\right] - \left[\frac{3}{4} - 1\right]$$

$$= [12 - 8] - \left[-\frac{1}{4}\right] = 4\frac{1}{4}$$

c) $$\int_{-1}^2 (x - 2)^2 dx = \int_{-1}^2 x^2 - 4x + 4dx = \left[\frac{x^3}{3} - 2x^2 + 4x\right]_{-1}^2 =$$

$$\left[\frac{8}{3} - 2^3 + 8\right] - \left[\frac{(-1)^3}{3} - 2(-1)^2 + 4(-1)\right] = \left[\frac{8}{3}\right] - \left[-6\frac{1}{3}\right] = 9$$

Method notes

For definite integrals

1. Integrate using the rule of integration, writing the answer inside square brackets and the upper and lower limits at the top and bottom of these brackets respectively
2. Evaluate the expression inside the brackets using the upper limit value for x.
3. Evaluate the expression inside the brackets using the lower limit value for x.
4. Subtract the lower limit evaluation from the upper limit evaluation.

Indefinite integration

In Core 1 we also looked at integration as the reverse process of differentiation.

When we do not have upper and lower limits for the integral we use the following rule:

if $\frac{dy}{dx} = x^n$ then $y = \int x^n dx = \frac{x^{n+1}}{n+1} + c$ $n \neq -1$ and c is a constant.

This is called **indefinite integration** since it does not have a finite value.

Method notes

You need to change $\sqrt{x}$ and $\frac{1}{x^2}$ into the form x^n using the rules of indices.

Example

Evaluate the indefinite integral $\int\left(2x - 3\sqrt{x} + \frac{1}{x^2}\right)\mathrm{d}x$.

Answer

Step 1: Rewrite: $\int\left(2x - 3\sqrt{x} + \frac{1}{x^2}\right)\mathrm{d}x = \int(2x - 3x^{\frac{1}{2}} + x^{-2})\mathrm{d}x$

Step 2: Apply the rule of integration: $x^2 - 3\frac{x^{\frac{3}{2}}}{\frac{3}{2}} + \frac{x^{-1}}{-1} + c$

Step 3: Simplify: $x^2 - 2x^{\frac{3}{2}} - \frac{1}{x} + c$

Area between a curve and the x-axis

Curves above the x-axis

We can now extend the ideas above of finding the area between a straight line graph and the x-axis to the area under any curve and the x-axis.

Example

Find the area of the region bounded by the curve with equation $y = x^2 - 5x + 10$, the x-axis and the lines $x = 2$ and $x = 6$

Answer

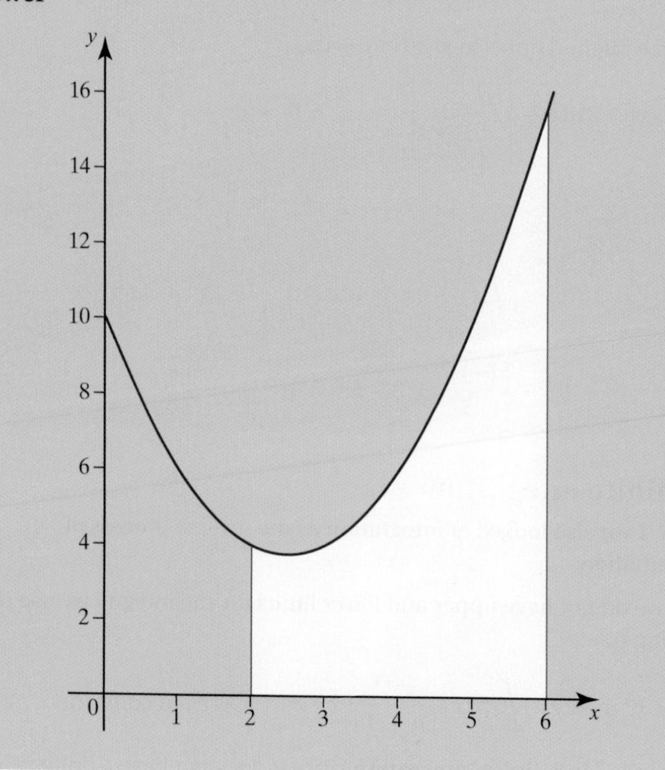

Fig. 7.2
The area of the region bounded by the curve with equation $y = x^2 - 5x + 10$, the x-axis and the lines $x = 2$ and $x = 6$.

Method notes

Sketch the curve and the lines to show the required region.

The area of the region is then the definite integral between the limits $x = 2$ and $x = 6$

$$\text{Area} = \int_2^6 (x^2 - 5x + 10)\,dx$$

$$= \left[\frac{x^3}{3} - 5\frac{x^2}{2} + 10x\right]_2^6$$

$$= \left[\frac{6^3}{3} - \frac{5 \times 6^2}{2} + 60\right] - \left[\frac{2^3}{3} - \frac{5 \times 2^2}{2} + 20\right]$$

$$= 42 - 12\frac{2}{3} = 29\frac{1}{3}$$

Stop and think 1

1. a) *Find the area bounded by the curve $y = x^2$, the x-axis and the lines $x = -3$ and $x = 3$*

   b) *Find the finite area bounded by the curve $y = x^2$ and the line $y = 9$*

Curves under the x-axis

Great care must be taken when finding an area between the x-axis and a curve which lies below the x-axis or straddles the x-axis if part of the curve is above the x-axis and part is below the x-axis as indicated in the graphs below.

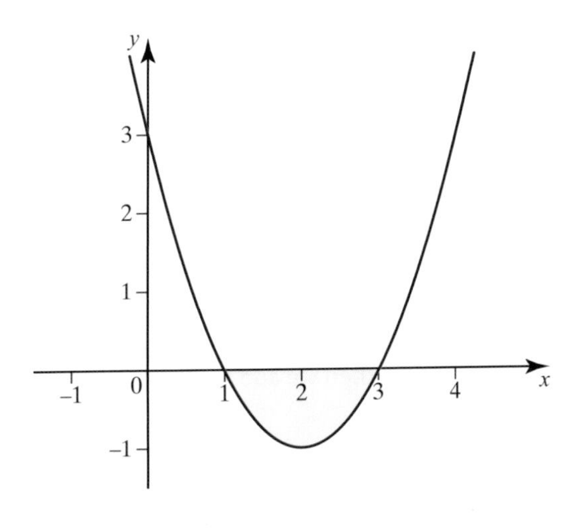

Fig. 7.3
The area between the graph of $y = (x-1)(x-3)$ and the x-axis.

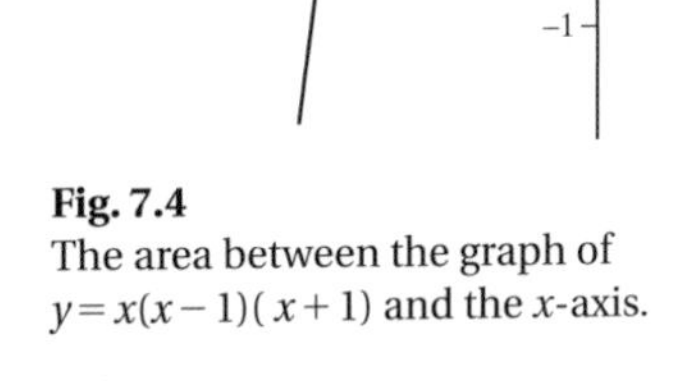

Fig. 7.4
The area between the graph of $y = x(x-1)(x+1)$ and the x-axis.

Method notes

Multiply out $(x-1)(x-3)$ before using the rule of integration.

To find the finite area of the graph below the x-axis in Figure 7.3 we use the definite integral of $y=(x-1)(x-3)$ between the limits $x=1$ and $x=3$.

$$\int_1^3 (x-1)(x-3)\,dx = \int_1^3 (x^2-4x+3)\,dx$$

$$=\left[\frac{x^3}{3}-2x^2+3x\right]_1^3 = [9-18+9]-\left[\frac{1}{3}-2+3\right]=-1\frac{1}{3}$$

Essential notes

A function is negative when its graph is below the x-axis.

The value of the definite integral is negative because the y-values in the interval (1, 3) are negative. However, this area has a numerical value therefore it is a positive quantity. We therefore deduce that the shaded area in Figure 7.3 is $1\frac{1}{3}$.

Generally if a function $f(x)$ is negative for all values of x in the interval from $x=a$ to $x=b$ then the area bounded by the curve $y=f(x)$, the x-axis and the lines $x=a$ and $x=b$ is:

$$\mathbf{A}=-\int_a^b f(x)\,dx$$

Method notes

Multiply out $x(x-1)(x+1)$ before using the rule of integration.

Using $x=-1$ and $x=0$ as the limits gives the area above the axis as $\frac{1}{4}$.

The area below the axis is exactly the same size as can be seen from the symmetry of the graph which indicates that it has the same numerical value as the area above the axis.

For the graph in Figure 7.4 it is tempting to think that the required finite area is the definite integral of $y=x(x-1)(x+1)$ between the limits -1 and $+1$. However, this would give:

$$\int_{-1}^1 x(x-1)(x+1)\,dx = \int_{-1}^1 (x^3-x)\,dx$$

$$=\left[\frac{x^4}{4}-\frac{x^2}{2}\right]_{-1}^1$$

$$=\left[\frac{1^4}{4}-\frac{1^2}{2}\right]-\left[\frac{(-1)^4}{4}-\frac{(-1)^2}{2}\right]$$

$$=0$$

Clearly the area of the shaded area is not zero! In this case each area is the same size which is why we have the answer 0.

To find the area we calculate the areas of the two regions separately using -1 and 0 and then 0 and 1 as our limits and then we add together the numerical values which gives an area $=\frac{1}{2}$.

Method notes

Always simplify the algebra of any function to be integrated before using the rule of integration.

Example

Find the area of the region bounded by the curve with equation $y=x(x-3)(x+1)$ and the x-axis.

Answer

Step 1: The curve intersects the x-axis when $y=0$ so $0=x(x-3)(x+1)$ so $x=0$, $x=-1$, $x=3$ which are the x limit values to use for definite integration.

Step 2: A sketch of the curve shows the required areas will be found by calculating the area above and below the axis separately.

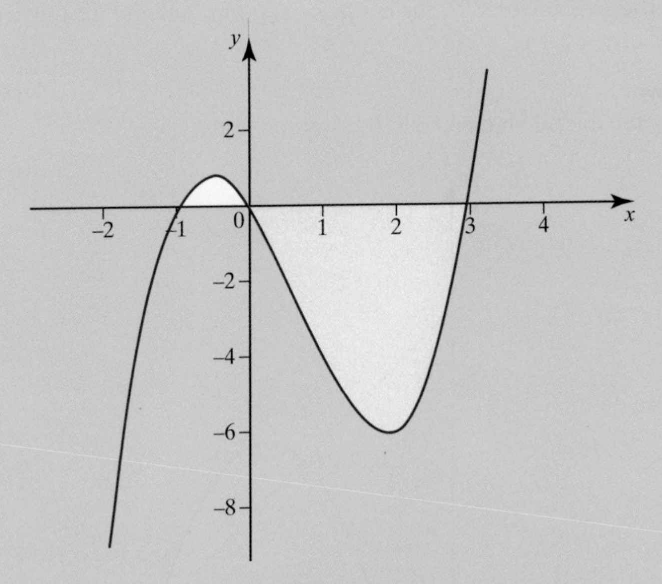

Fig. 7.5
The graph of equation $y = x(x - 3)\ (x + 1)$.

Step 3: Area A_1 between $x = -1$ and $x = 0$ is above the axis so:

$$A_1 = \int_{-1}^{0} x(x+1)(x-3)\mathrm{d}x$$

$$= \int_{-1}^{0} (x^3 - 2x^2 - 3x)\mathrm{d}x$$

$$= \left[\frac{x^4}{4} - \frac{2x^3}{3} - \frac{3x^2}{2}\right]_{-1}^{0}$$

$$= [0] - \left[\frac{(-1)^4}{4} - \frac{2(-1)^3}{3} - \frac{3(-1)^2}{2}\right]$$

$$= \frac{7}{12}$$

Step 4: Area A_2 between $x = 0$ and $x = 3$ is below the x-axis so

$$A_2 = -\int_{0}^{3} x(x+1)(x-3)\mathrm{d}x = -\int_{0}^{3} (x^3 - 2x^2 - 3x)\mathrm{d}x$$

$$= -\left[\frac{x^4}{4} - \frac{2x^3}{3} - \frac{3x^2}{2}\right]_{0}^{3} = -\left(\left[\frac{3^4}{4} - \frac{2 \times 3^3}{3} - \frac{3 \times 3^2}{2}\right]_{0}^{3} - 0\right)$$

$$= -\left(-11\frac{1}{4}\right) = 11\frac{1}{4}$$

Area of the shaded region $= A_1 + A_2 = \frac{7}{12} + 11\frac{1}{4} = 11\frac{5}{6}$

Area between a curve and a line

Example

Find the area enclosed by the curve $y=(1-x)(x-4)$ and the line with equation $x+y=4$

Answer

Using the method indicated in the Method notes:

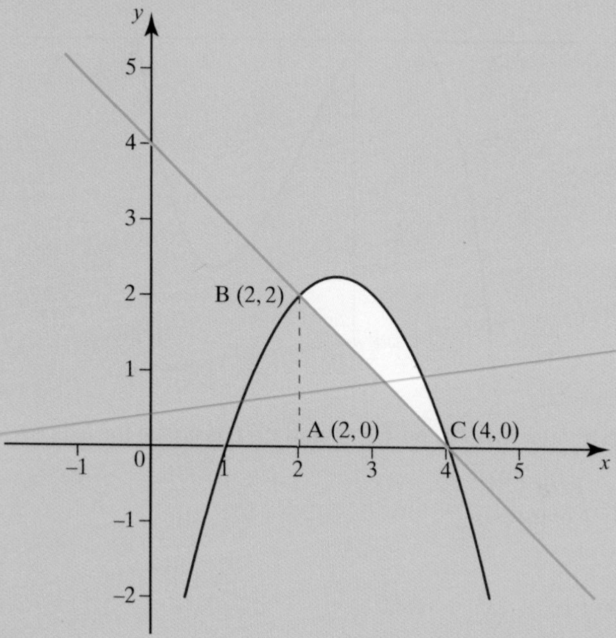

Fig. 7.6
The graph of equations $y=(1-x)(x-4)$ and $x+y=4$.

Method notes

When finding areas between lines and curves always:

1. Sketch the graphs and shade the required area.
2. Find where the line and the curve intersect to find the x limit values to use for definite integration.
3. Write down the equation of the curve which forms the upper boundary of the required area.
4. The required area will be definite integral using the upper boundary equation minus the definite integral using the lower boundary equation.

The line and curve intersect where $(1-x)(x-4)=4-x$

$\Rightarrow$ $$(1-x)(x-4)+(x-4)=0$$

$\Rightarrow$ $$(2-x)(x-4)=0$$

therefore $x=2$ or $x=4$ are the x limits for definite integration.

$$\text{The required area} = \underbrace{\int_2^4 (1-x)(x-4)\,dx}_{\text{shaded area} + \Delta ABC} - \underbrace{\int_2^4 (4-x)\,dx}_{\text{area } \Delta ABC}$$

$$\text{Area} = \int_2^4 (1-x)(x-4)-(4-x)\,dx = \int_2^4 (-4+5x-x^2)-(4-x)\,dx$$

$$= \int_2^4 (-8+6x-x^2)\,dx = \left[-8x+3x^2-\frac{x^3}{3}\right]_2^4$$

$$= \left[-32+48-\frac{64}{3}\right]-\left[-16+12-\frac{8}{3}\right]=1\frac{1}{3}$$

In the previous example the area of the shaded region was found by calculating the integral $\int_a^b (y_1 - y_2)\mathrm{d}x$ where $y_1 = (1 - x)(x - 4)$ was the 'upper' function and $y_2 = 4 - x$ was the 'lower' function.

General formula

The area between a curve ('upper' function y_1) and a line ('lower' function y_2) is given by $\int_a^b (y_1 - y_2)\mathrm{d}x$ where $x = a$ and $x = b$ are the x-coordinates of the points of intersection of the line and the curve.

The formula $\int_a^b (y_1 - y_2)\mathrm{d}x$ can also be used when part of the region is below the x-axis.

Method notes

Since the limits are the same we can combine the functions before integrating.

Essential notes

You should learn this general formula for the examination.

Essential notes

y_1 represents the upper boundary.

y_2 represents the lower boundary.

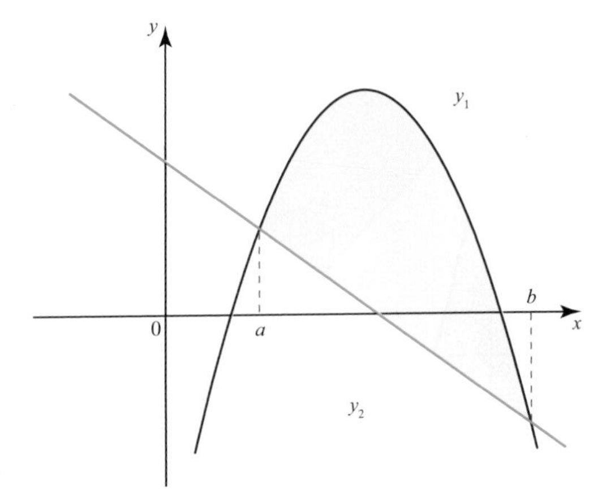

Fig. 7.7
Graphs with part of the region below the x-axis.

This can be demonstrated by moving the two graphs upwards so that they are both above the x-axis. This does not change the limits or the size of the area of the region as shown in the graph below and the area of the shaded region is given by:

$$\int_a^b ((y_1 + c) - (y_2 + c))\mathrm{d}x = \int_a^b (y_1 - y_2)\mathrm{d}x$$

Exam tips

You do not need to translate the graphs when finding areas in examination questions. This is only to show the general result.

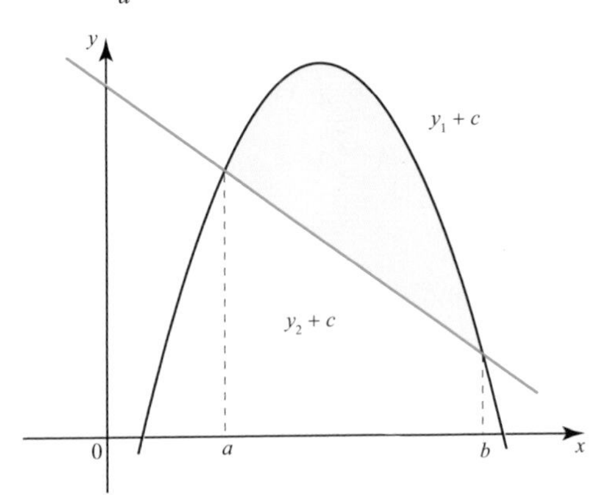

Fig. 7.8
Graphs from figure 7.7 with the region moved above the x-axis.

Method notes

1. Find the x-coordinates of the points of intersection.
2. These are the limits for the definite integration: $\int_a^b (y_1 - y_2)\mathrm{d}x$

Example

Find the area enclosed by the curve $y = x^2 - 5x + 3$ and the line with equation $y = 3 - x$ as shown in Figure 7.9 below.

Answer

The line and the curve intersect where $x^2 - 5x + 3 = 3 - x$

$\Rightarrow x^2 - 4x = 0$

$\Rightarrow x = 0$ or $x = 4$

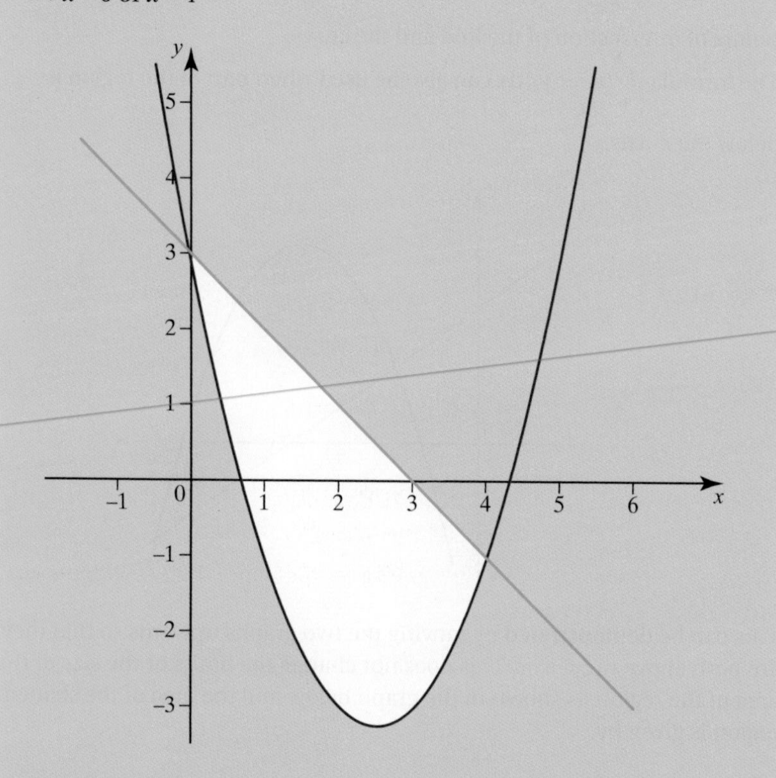

Fig. 7.9
The graph of equations $y^2 = x^2 - 5x + 3$ and $y = 3 - x$

Area of the shaded region $= \int_0^4 ((3 - x) - (x^2 - 5x + 3))\mathrm{d}x$

Simplify the algebra to give $\int_0^4 (4x - x^2)\mathrm{d}x = \left[2x^2 - \frac{x^3}{3}\right]_0^4$

$$= \left[2 \times 4^2 - \frac{4^3}{3}\right]_0^4 - [0]$$

$$= 10\frac{2}{3}$$

The trapezium rule

At this stage of your mathematics course you cannot evaluate integrals such as $\int_1^2 3^x \mathrm{d}x$ and $\int_0^2 \sqrt{2x-1}\,\mathrm{d}x$.

However, the link between definite integrals and areas provides a method of approach that allows us to find an approximation to the value of integrals like these. The method is called the **trapezium rule.**

Consider the integral $\int_1^2 3^x \mathrm{d}x$. This is the area between the curve $y = 3^x$, the x-axis and the ordinates $x = 1$ and $x = 2$ shown in Figure 7.10.

Essential notes

Exact evaluation of integrals such as $\int_1^2 3^x \mathrm{d}x$ and $\int_0^2 \sqrt{2x-1}\,\mathrm{d}x$ will be covered in Core 3 and Core 4.

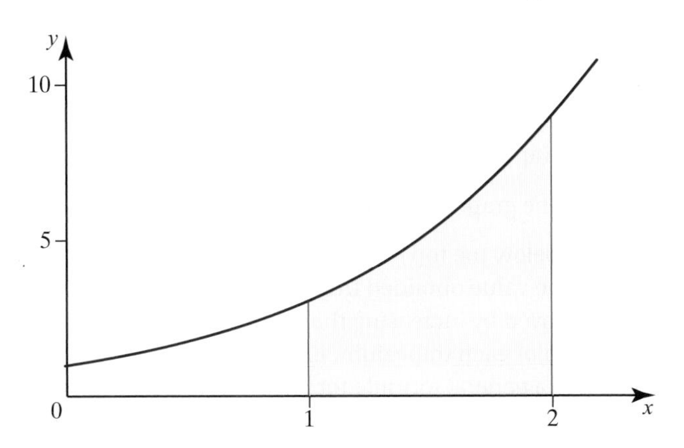

Fig. 7.10
$\int_1^2 3^x \mathrm{d}x$ as an area.

This approximation method involves dividing the area up into equal strips and drawing trapezia to approximate the required area.

Suppose that we use **four** strips as shown in Figure 7.11 below then each strip has width 0.25.

Essential notes

Strip width is (upper limit – lower limit) which is then divided by the number of strips, in this case $\frac{2-1}{4} = 0.25$

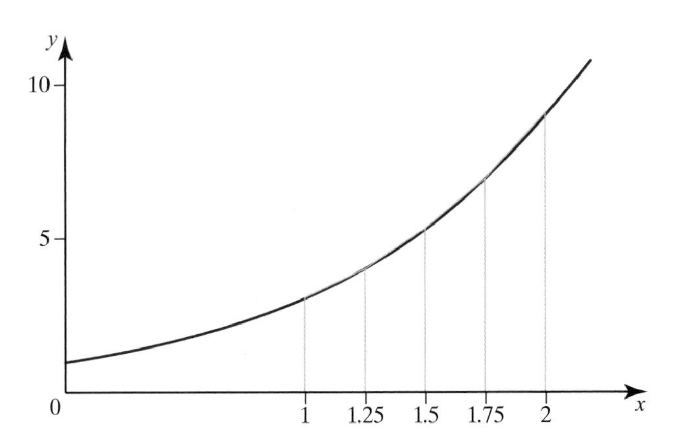

Fig. 7.11
The region approximated by 4 trapezia.

The equation of the curve is $y = 3^x$ and the heights of the trapezia in Figure 7.11 are the values of $y = 3^x$ at $x = 1$, $x = 1.25$, $x = 1.5$, $x = 1.75$, and

$x = 2$ respectively. Each trapezium has a width of 0.25 so the total area of the four trapezia is:

$$\frac{1}{2}(3^1 + 3^{1.25}) \times 0.25$$
$$+\frac{1}{2}(3^{1.25} + 3^{1.5}) \times 0.25$$
$$+\frac{1}{2}(3^{1.5} + 3^{1.75}) \times 0.25$$
$$+\frac{1}{2}(3^{1.75} + 3^2) \times 0.25$$

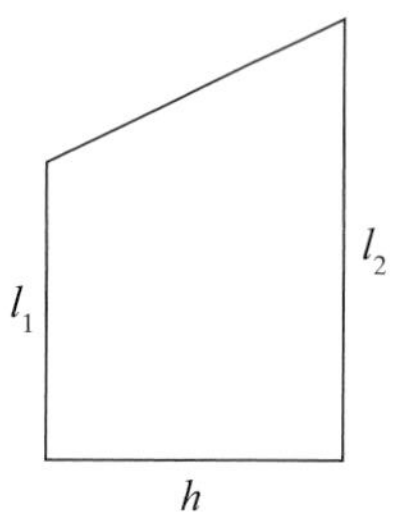

Fig. 7.12

Simplifying this expression gives:

$$\frac{1}{2} \times 0.25(3^1 + 2(3^{1.25} + 3^{1.5} + 3^{1.75}) + 3^2) = 5.496$$

The actual value is $\int_1^2 3^x \mathrm{d}x = 5.461$ which will be covered in Core 4 but the answer using the trapezium rule gives a good approximation.

You can see from the graph of $\int_1^2 3^x \mathrm{d}x$ and the four trapezia that in each strip the graph is below the top of the trapezium so that the required area will be less than the value obtained from the trapezium rule. The accuracy can often be improved by increasing the number of strips and hence reducing the width of each trapezium. Extending this idea gives the **trapezium rule** as a general formula for approximating an integral:

$$\int_a^b y\mathrm{d}x \approx \tfrac{1}{2}h[y_0 + 2(y_1 + y_2 + y_3 + \ldots + y_{n-2} + y_{n-1}) + y_n]$$

where the strip width is $h = \dfrac{b-a}{n}$, n is the number of strips and $y_0 = \mathrm{f}(a)$, $y_1 = \mathrm{f}(a+h)$, $y_2 = \mathrm{f}(a+2h)$ and $y_i = \mathrm{f}(a+ih)$.

Essential notes

The area of a trapezium is $\frac{1}{2}(l_1 + l_2) \times h$ where l_1 and l_2 are the lengths of the parallel sides and h is the width of the trapezium as shown in Figure 7.12.

Essential notes

Note that this formula adds together and doubles the y values which appear twice in the sum then adds on the first and last y values. This complete sum is then multiplied by half of the strip width value $\left(\frac{1}{2} \times 0.25\right)$.

Examination tips

You do not need to learn this formula as it is in your formula booklet.

You do need to learn how to apply it!

Example

Use the trapezium rule with 4 strips to estimate the value of:

$$\int_1^3 \frac{1}{x^2+1}\mathrm{d}x.$$

Answer

Step 1: State the value of n. In this question $n = 4$

Step 2: State the a and b values. In this question $a = 1$ $b = 3$

Step 3: Work out the strip width using the formula $h = \dfrac{b-a}{n}$

$$\Rightarrow h = \frac{3-1}{4} = \frac{1}{2}$$

Step 4: For each x-value work out the corresponding y-value and set out the results in a table.

In this question $y = 1/(x^2 + 1)$ so

x	1.0	1.5	2	2.5	3.0
y	0.5	0.30769	0.2	0.13793	0.1

Step 5: Use the trapezium rule from the formula booklet to find the area of the 4 trapezia.

$$= \frac{1}{2} \times \underset{\substack{\uparrow \\ h}}{\frac{1}{2}} \times [\underset{\substack{\uparrow \\ y_0}}{0.5} + 2(\underset{\substack{\uparrow \\ y_1}}{0.30769} + \underset{\substack{\uparrow \\ y_2}}{0.2} + \underset{\substack{\uparrow \\ y_3}}{0.13793}) + \underset{\substack{\uparrow \\ y_4}}{0.1}] = 0.473$$

Therefore:

$$\int_1^3 \frac{1}{x^2 + 1}\,\mathrm{d}x \approx 0.473$$

Method notes

Setting the results out as a table helps when working with the trapezium rule.

Stop and think answers

1. *The curve $y = x^2$ is symmetrical about the y-axis, passes through the origin (0, 0) and the curve is always above the x-axis.*
   The boundary values for x are -3 and $+3$.

$$\text{Area} = \int_{-3}^{3} y\,\mathrm{d}x \text{ so } \int_{-3}^{3} x^2\,\mathrm{d}x = \left[\frac{x^3}{3}\right]_{-3}^{3} = \left[\frac{27}{3} - -\frac{27}{3}\right] = 18$$

Questions

You may use a calculator.

A formula sheet is attached for your reference.

1. a) Find the remainder when

   $x^3 - 6x^2 + 3x + 10$

   is divided by

   i) $x - 1$

   ii) $x + 1$ (3)

   b) Hence, or otherwise, find all the solutions to the equation

   $x^3 - 6x^2 + 3x + 10 = 0$ (4)

2. A geometric series has first term 6 and common ratio $\frac{2}{3}$. Calculate:

   a) the 10th term of the series, to 3 decimal places (2)

   b) the sum to infinity of the series. (2)

   Given that the sum to n terms of the series is greater than 17.5

   c) find the least value of n. (5)

3. a) Find the first 3 terms, in ascending powers of x, of the binomial expansion of $(1 + px)^{10}$ where p is a constant. (4)

   b) The first 3 terms are 1, $30x$, $27qx^2$ where q is a constant. Find the values of p and q. (2)

4. a) Show that the equation

   $3\cos^2\theta - 2\sin^2\theta = 1$

   can be written as

   $5\cos^2\theta = 3$ (2)

   b) Hence solve the equation $3\cos^2\theta - 2\sin^2\theta = 1$ for $0° \leq \theta \leq 360°$ giving your answers to 1 decimal place. (7)

5. a) Find, to 3 significant figures, the value of x for which $3^x = 5$ (2)

   b) Solve the equation $3^{2x} - 9(3^x) + 20 = 0$ (5)

6. a) If $y = 5^x$ complete this table giving the values of 5^x to 3 decimal places. (2)

x	0	0.2	0.4	0.6	0.8	1
y	1		1.904			5

   b) Use the trapezium rule with 5 strips to find an approximation for the value of

   $$\int_0^1 5^x \mathrm{d}x$$ (4)

7. The figure shows the triangle ABC, in which $AB = 3$ cm, $BC = 5$ cm and the angle $ABC = 2.1$ radians.

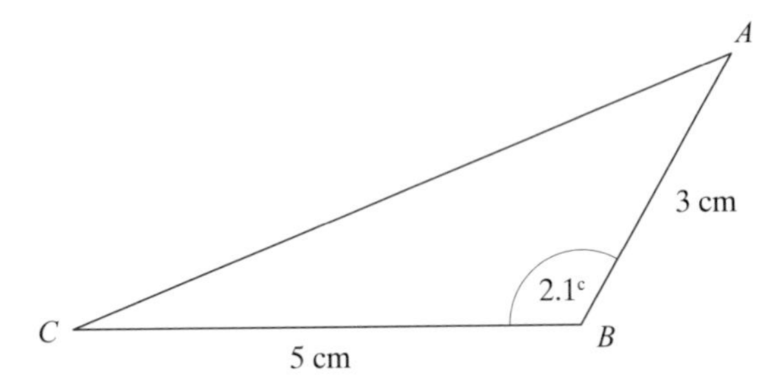

Calculate:

a) the length of the side AC, giving your answer to 2 decimal places. (3)

b) the angle BAC, giving your answer in radians to 2 decimal places. (3)

c) the area of the triangle ABC. (2)

8. In the figure the curve C has equation $y = 5x - x^2$ and the line L has equation $y = 2x$.

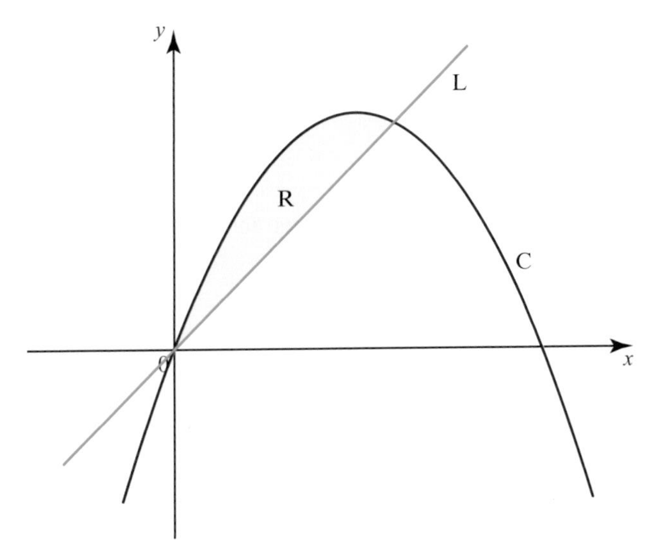

a) Find the values of x where C intersects the x axis. (1)

b) Show that the line L intersects the curve C at the points (0, 0) and (3, 6). (3)

c) Find the area of the shaded region R, bounded by the curve C and the line L. (4)

9. a) Find the coordinates of the stationary points of the curve with equation

$y = x^3 - 8x^2 + 20x$ (4)

b) Find $\frac{d^2y}{dx^2}$ and use your answer to classify the stationary points. (4)

10. A circle C has centre (5, 3) and passes through the point P(8, 5).

  a) Find the equation for C. (3)

  b) Find the equation of the tangent to C at point P, giving your answer in the form $ax + by + c = 0$ where a, b and c are integers. (4)

Answers

1. a) $f(x) = x^3 - 6x^2 + 3x + 10$

  Use the remainder theorem to find a remainder. (1)

  i) when $f(x)$ is divided by $(x - 1)$ put $x - 1 = 0$ so $x = 1$ then evaluate $f(1)$ to give the remainder.

  $f(1) = 1^3 - 6(1^2) + 3(1) + 10 = 8$ so the remainder is 8 (1)

  ii) put $x + 1 = 0$ so $x = -1$ then evaluate $f(-1)$

  $f(-1) = (-1)^3 - 6(-1)^2 + 3(-1) + 10 = -1 - 6 - 3 + 10 = 0$ so the remainder is 0 (1)

  b) Hence means there is a connection to part (a) of the question.

  In part ii) the remainder = 0 when $f(x)$ is divided by $(x + 1)$ so using the factor theorem we know that $(x + 1)$ is a factor of $f(x)$

  $x^3 - 6x^2 + 3x + 10 = 0$

  $\Rightarrow (x + 1)(x^2 - 7x + 10) = 0$ (1)

  Factorising again:

  $(x + 1)(x - 2)(x - 5) = 0$ (2)

  so $x = -1$, $x = 2$, $x = 5$ are the three solutions of the cubic equation. (1)

2. Geometric series with $a = 6$, $r = \frac{2}{3}$

  a) 10th term $= ar^9 = 6 \times \left(\frac{2}{3}\right)^9 = 0.156$ (2)

  b) Sum to infinity $= \frac{a}{(1 - r)} = \frac{6}{1 - \frac{2}{3}} = 18$ (2)

  c) Sum of n terms is > 17.5

  $\Rightarrow S_n = \frac{a(1 - r^n)}{1 - r}$ (as $r < 1$) (1)

  $$\Rightarrow \frac{6\left(1 - \left(\frac{2}{3}\right)^n\right)}{1 - \frac{2}{3}} > 17.5$$

  $$\Rightarrow 1 - \left(\frac{2}{3}\right)^n > \frac{17.5}{18}$$

  $$\Rightarrow 0.0277 > \left(\frac{2}{3}\right)^n$$ (1)

  Taking logs of both sides gives $\log 0.0277 > n \log 0.6666$

Divide by log 0.6666 which is negative so reverse the inequality

$\Rightarrow \dfrac{\log 0.0277}{\log 0.6666} < n$ (2)

$\Rightarrow 8.838 < n$ so least value of n is 9 (1)

3. a) Binomial expansion of $(1+px)^{10} = 1 + 10(px)^1 + \dfrac{10(9)(px)^2}{2.1}$ so

the first three terms are:

$1 + 10\,px + 45p^2x^2$ (4)

b) Given that the first three terms are 1, $30x$, $27qx^2$:

$10px = 30x$ and $45p^2x^2 = 27qx^2$

$10p = 30$ so $p = 3$ and $45p^2 = 27q$ so $45 \times 9 = 27q$ so $15 = q$ (2)

4. a) Given $3\cos^2\theta - 2\sin^2\theta = 1$:

Use the Pythagorean identity with $\sin^2\theta = 1 - \cos^2\theta$ and rewrite the original equation as:

$3\cos^2\theta - 2(1 - \cos^2\theta) = 1$

$\Rightarrow 3\cos^2\theta - 2 + 2\cos^2\theta = 1$

$\Rightarrow 5\cos^2\theta = 3$ as required. (2)

b) Hence implies that the result in part (a) of the question will be useful Use $5\cos^2\theta = 3$ instead of the given equation as it is easier to solve:

$5\cos^2\theta = 3$

$\Rightarrow \cos^2\theta = \dfrac{3}{5}$

$\Rightarrow \cos\theta = \sqrt{0.6}$ or $\cos\theta = -\sqrt{0.6}$ (3)

Using a calculator in degrees mode and the CAST diagram for $0° \leq \theta \leq 360°$

$\theta = 39.23°, 360° - 39.23°, 140.78°, 180° + 39.23°$

$\theta = 39.2°, 320.8, 140.8°, 219.2°$ (4)

5. a) Given $3^x = 5$ take logs of both sides: $\log 3^x = \log 5$

Using the rules of logarithms:

$x \log 3 = \log 5$

$x = \dfrac{\log 5}{\log 3}$

$= 1.46$ (3 s.f.) (2)

b) To solve $3^{2x} - 9(3^x) + 20 = 0$ you need to recognise that, by the rules of indices, $3^{2x} = (3^x)^2$ (1)

So the original equation is $(3^x)^2 - 9(3^x) + 20 = 0$

Let $3^x = X$ then the equation is $X^2 - 9X + 20 = 0$

Factorising this quadratic equation gives $(X-4)(X-5)=0$ (2)

$\Rightarrow X=4, X=5$

$\Rightarrow 3^x=4 \quad 3^x=5$

Take logs and use the rules of logs to solve $3^x=4$

$\Rightarrow x=\dfrac{\log 4}{\log 3}=1.26$ (1)

We have already solved $3^x=5$ in part (a) where $x=1.46$ (1)

6. a) Given $y=5^x$:

$x=0.2, y=1.379$

$x=0.6, y=2.627$

$x=0.8, y=3.624$ (2)

b) 5 strips so 6 ordinates (or y values)

width of each strip $=\dfrac{1-0}{5}=0.2$

approximation using the trapezium rule:

$=\dfrac{1}{2}\times 0.2[1+2(1.379+1.904+2.627+3.624)+5]=2.507$ (4)

7. Use conventional labelling ($a=5$, $c=3$, $B=2.1^c$) and the calculator in RAD mode throughout this question.

a) Two sides and the angle between them are given so use the cosine rule:

$b^2=a^2+c^2-2ac\cos B=25+9-2(5)(3)\cos 2.1^c$

$b^2=34+15.1453$ so $b=7.01$(2 d.p) (3)

b) $\angle BAC$ is $\angle A$ using conventional labelling. Use the rewrite of the cosine rule:

$\cos A=\dfrac{b^2+c^2-a^2}{2bc}$

$\Rightarrow \cos A=\dfrac{49.145+9-25}{2(7.01)(3)}$

$\Rightarrow \quad \angle A=0.66$ radians (3)

c) area of triangle $ABC=\dfrac{1}{2}ac\sin B=\dfrac{1}{2}\times 5\times 3\times \sin 2.1^c=7.5\times 0.8632$

so area $ABC=6.47$ (2)

8. Curve C equation is $y=5x-x^2$ and line L equation is $y=2x$.

a) C intersects the x axis when $y=0$

so $0=5x-x^2$

Factorising gives

$0=x(5-x)$

so $x=0$ or $x=5$ (1)

b) Where C and L intersect the y values on line and curve are equal

$$2x = 5x - x^2$$

$$x^2 = 3x$$

$$x(x - 3) = 0$$

so $x = 0, x = 3$

when $x = 0, y = 2(0)$ so (0, 0) is one intersection point.

when $x = 3, y = 2(3)$ so (3, 6) is the other intersection point. (3)

c) Area of region R = $\int$[y (for curve) – y (for line)] dx between the x limits of 3 and 0 (the x intersection values from part b)

so $\int_0^3 5x - x^2 - (2x)\,dx = \int_0^3 3x - x^2\,dx$ so $\left[\frac{3x^2}{2} - \frac{x^3}{3}\right]$ (3)

so $\frac{3(9)}{2} - \frac{3^3}{3} - (0)$ and area of region = 4.5 square units (1)

9. a) Stationary points occur where $\frac{dy}{dx} = 0$

$$y = x^3 - 8x^2 + 20x$$

so $\frac{dy}{dx} = 3x^2 - 16x + 20$

At the stationary points: $3x^2 - 16x + 20 = 0$

Factorising gives: $(3x - 10)(x - 2) = 0$

so $x = \frac{10}{3}, x = 2$

when $x = 2, y = 16$ which gives (2, 16) as a stationary point. (2)

The other stationary point is when $x = \frac{10}{3}$ so y = 14.84 which gives $\left(\frac{10}{3}, 14.84\right)$ as the other stationary point. (2)

b) $\frac{dy}{dx} = 3x^2 - 16x + 20$ so $\frac{d^2y}{dx^2} = 6x - 16$ (1)

The sign of the second derivative classifies stationary points easily.

If $\frac{d^2y}{dx^2}$ is positive at any point there is a minimum turning point at that point.

If $\frac{d^2y}{dx^2}$ is negative at any point there is a maximum turning point at that point.

If $x = 2$, $\frac{d^2y}{dx^2} = 6x - 16 = 6(2) - 16$ which is negative therefore there is a maximum turning point at (2, 16)

If $x = \frac{10}{3}$, $\frac{d^2y}{dx^2} = 6x - 16 = 6\left(\frac{10}{3}\right) - 16$ which is positive therefore there is a minimum turning point at $\left(\frac{10}{3}, 14.84\right)$ (3)

10. Circle has centre (5, 3) and passes through P(8, 5)

a) We can use the form $(x - a)^2 + (y - b)^2 = r^2$ as we know $a = 5$, $b = 3$

We also know that the radius $r =$ distance from centre to point P as we were told P was a point on the circle.

Using straight line geometry:

$r = \sqrt{(x_2 - x_1)^2 + (y_2 - y_1)^2}$ where $x_1 = 5$, $y_1 = 3$, $x_2 = 8$, $y_2 = 5$

so $r = \sqrt{3^2 + 2^2}$

$r = \sqrt{13}$

So the equation of the circle is $(x - 5)^2 + (y - 3)^2 = 13$ (3)

b) The tangent at P(8, 5) is at 90° to the radius.

If point Q(5, 3) is the centre of the circle then

gradient QP $= \frac{5 - 3}{8 - 5} = \frac{2}{3}$ so gradient of tangent at P $= -\frac{3}{2}$ (since product of gradients of perpendicular lines $= -1$) (1)

Use $y - y_1 = m(x - x_1)$ to find the equation of the tangent at P(8, 5) with $x_1 = 8$, $y_1 = 5$, $m = -\frac{3}{2}$

$\Rightarrow y - 5 = -\frac{3}{2}(x - 8)$

$\Rightarrow \quad y = -\frac{3}{2}x + 12 + 5$

So the equation is $0 = -3x - 2y + 34$ in the required form. (3)

Set notation

$\in$	is an element of
$\notin$	is not an element of
$\{x_1, x_2, \ldots\}$	the set with elements $x_1, x_2, \ldots$
$\{x: \ldots\}$	the set of all x such that ...
$n(A)$	the number of elements in set A
$\varnothing$	the empty set
ε	the universal set
A'	the complement of the set A
$\mathbb{N}$	the set of natural numbers, $\{1, 2, 3, \ldots\}$
$\mathbb{Z}$	the set of integers, $\{0, \pm1, \pm2, \pm3, \ldots\}$
$\mathbb{Z}^+$	the set of positive integers, $\{1, 2, 3, \ldots\}$
$\mathbb{Z}_n$	the set of integers modulo n, $\{0, 1, 2, \ldots, n-1\}$
$\mathbb{Q}$	the set of rational numbers, $\left\{\frac{p}{q} : p \in \mathbb{Z}, q \in \mathbb{Z}^+\right\}$
$\mathbb{Q}^+$	the set of positive rational numbers, $\{x \in \mathbb{Q} : x > 0\}$
$\mathbb{Q}_0^+$	the set of positive rational numbers and zero, $\{x \in \mathbb{Q} : x \geq 0\}$
$\mathbb{R}$	the set of real numbers
$\mathbb{R}^+$	the set of positive real numbers $\{x \in \mathbb{R} : x > 0\}$
$\mathbb{R}_0^+$	the set of positive real numbers and zero, $\{x \in \mathbb{R} : x \geq 0\}$
$\mathbb{C}$	the set of complex numbers
(x, y)	the ordered pair x, y
$A \times B$	the cartesian product of sets A and B, ie $A \times B = \{(a, b) : a \in A, b \in B\}$
$\subseteq$	is a subset of
$\subset$	is a proper subset of
$\cup$	union
$\cap$	intersection
$[a, b]$	the closed interval, $\{x \in \mathbb{R} : a \leq x \leq b\}$
$[a, b), [a, b[$	the interval $\{x \in \mathbb{R} : a \leq x < b\}$
$(a, b],]a, b]$	the interval $\{x \in \mathbb{R} : a < x \leq b\}$
$(a, b),]a, b[$	the open interval $\{x \in \mathbb{R} : a < x < b\}$
yRx	y is related to x by the relation R
$y \sim x$	y is equivalent to x, in the context of some equivalence relation

Miscellaneous symbols

$=$	is equal to
$\neq$	is not equal to
$\equiv$	is identical to or is congruent to
$\approx$	is approximately equal to
$\cong$	is isomorphic to
$\propto$	is proportional to
$<$	is less than
$\leq$	is less than or equal to, is not greater than
$>$	is greater than
$\geq$	is greater than or equal to, is not less than
∞	infinity
$p \wedge q$	p and q
$p \vee q$	p or q (or both)
$\sim p$	not p
$p \Rightarrow q$	p implies q (if p then q)
$p \Leftarrow q$	p is implied by q (if q then p)
$p \Leftrightarrow q$	p implies and is implied by q (p is equivalent to q)
$\exists$	there exists
$\forall$	for all

Operations

$a + b$	a plus b
$a - b$	a minus b
$a \times b$, ab, $a.b$	a multiplied by b
$a \div b$, $\frac{a}{b}$, a/b	a divided by b
$\sum_{i=1}^{n} a_i$	$a_1 + a_2 + \ldots + a_n$
$\prod_{i=1}^{n} a_i$	$a_1 \times a_2 \times \ldots \times a_n$
$\sqrt{a}$	the positive square root of a
$\lvert a \rvert$	the modulus of a
$n!$	n factorial
$\binom{n}{r}$	the binomial coefficient $\frac{n!}{r!(n-r)!}$ for $n \in \mathbb{Z}^+$ $\frac{n(n-1)\ldots(n-r+1)}{r!}$ for $n \in \mathbb{Q}$

Functions

$f(x)$	the value of the function f at x
$f: A \to B$	f is a function under which each element of set A has an image in set B
$f: x \to y$	the function f maps the element x to the element y
f^{-1}	the inverse function of the function f
$g \circ f$, gf	the composite function of f and g which is defined by $(g \circ f)(x)$ or $gf(x) = g(f(x))$
$\lim_{x \to a} f(x)$	the limit of $f(x)$ as x tends to a
Δx, dx	an increment of x
$\frac{dy}{dx}$	the derivative of y with respect to x
$\frac{d^n y}{dx^n}$	the nth derivative of y with respect to x
$f'(x), f''(x), \ldots, f^{(n)}(x)$	the first, second, ..., nth derivatives of $f(x)$ with respect to x the indefinite integral of y with respect to x
$\int y \, dx$	the indefinite integral of y with respect to x
$\int_a^b y \, dx$	the definite integral of y with respect to x between the limits $x = a$ and $x = b$
$\dot{x}, \ddot{x}, \ldots$	the first, second, ... derivatives of x with respect to t

Exponential and logarithmic functions

$\log_a x$	logarithm to the base a of x
$\lg x$, $\log_{10} x$	logarithm of x to base 10

Circular and hyperbolic functions

sin, cos, tan, cosec, sec, cot	the circular functions

Formulae you need to remember

Laws of logarithms

$\log_a x + \log_a y = \log_a (xy)$

$\log_a x - \log_a y \equiv \log_a \left(\frac{x}{y}\right)$

$k\log_a x = \log_a (x^k)$

Trigonometry

In the triangle ABC

$\frac{a}{\sin A} = \frac{b}{\sin B} = \frac{c}{\sin C}$

area $= \frac{1}{2} ab \sin C$

Area

area under a curve $= \int_a^b y \,\mathrm{d}x \ (y \geq 0)$

Formulae given in the formulae booklet

Mensuration

Surface area of sphere $= 4\pi r^2$

Area of curved surface of cone $= \pi r \times$ slant height

Arithmetic sequences

Formula for the nth term $u_n = a + (n-1)d$ where a is the first term and d is the common difference

Arithmetic series

$u_n = a + (n - 1)d$

$S_n = \frac{1}{2}n(a + l) = \frac{1}{2}n[2a + (n - 1)d]$

Cosine rule

$a^2 = b^2 + c^2 - 2bc \cos A$

Logarithms and exponentials

$\log_a x = \frac{\log_b x}{\log_b a}$

Binomial series

$$(a+b)^n = a^n + \binom{n}{1} a^{n-1}b + \binom{n}{2} a^{n-2}b^2 + \ldots + \binom{n}{r} a^{n-r}b^r + \ldots + b^n \quad (n \in \mathbb{N})$$

where $\binom{n}{r} = {}^nC_r = \dfrac{n!}{r!(n-r)!}$

$$(1+x)^n = 1 + nx + \frac{n(n-1)}{1 \times 2}x^2 + \ldots + \frac{n(n-1)\ldots(n-r+1)}{1 \times 2 \times \ldots \times r}x^r + \ldots \quad (|x| < 1, n \in \mathbb{R})$$

Geometric Series

$$u_n = ar^{n-1}$$

$$S_n = \frac{a(1-r^n)}{1-r}$$

$$S_\infty = \frac{a}{1-r} \text{ for } |r| < 1$$

Numerical integration

The trapezium rule: $\int_a^b y\,dx \approx \frac{1}{2}h\{(y_0 + y_n) + 2(y_1 + y_2 + \ldots + y_{n-1})\}$, where $h = \dfrac{b-a}{n}$

Index

G

I

L

M

O

P

Q

R

S